AP® BIOLOGY
CRASH COURSE®

Michael D'Alessio, M.S.

Lauren Gross, Ph.D.

Jennifer Guercio, M.S.

Christina Palffy, M.S.

Research & Education Association

www.rea.com

Research & Education Association
258 Prospect Plains Road
Cranbury, New Jersey 08512
Email: info@rea.com

AP® BIOLOGY CRASH COURSE, 3rd Edition

Printed in the United States of America

Library of Congress Control Number 2019949052

ISBN-13: 978-0-7386-1268-3
ISBN-10: 0-7386-1268-5

Cover image: © iStockphoto.com/dima_sidelnikov

AP® BIOLOGY CRASH COURSE TABLE OF CONTENTS

ABOUT OUR BOOK

REA's *AP® Biology Crash Course* is designed for the last-minute studier or any student who wants a quick refresher on the AP® course. This *Crash Course* is based on the latest changes to the AP® Biology course and exam and focuses only on the topics tested, so you can make the most of your study time.

Written by nationally recognized AP® Biology test experts, our *Crash Course* gives you a concise review of the major concepts and important topics tested on the AP® Biology exam.

- **Part I** offers you **Keys for Success**, so you can tackle the exam with confidence.

- **Part II** contains the **Content Review** chapters that cover every critical aspect of today's exam.

- **Part III** focuses on writing the essays for the AP® Biology exam and the science practices that will be tested, along with AP®-style practice questions to prepare you for what you'll see on the multiple-choice section of the exam.

- **Part IV** presents 13 inquiry-based lab experiments. These labs are included to build your critical-thinking powers so you can reason your way through real-world scenarios on the exam.

ABOUT OUR ONLINE PRACTICE EXAM

How ready are you for the AP® Biology exam? Find out by taking **REA's online practice exam** available at *www.rea.com/studycenter*. This test features automatic scoring, detailed explanations of all answers, and diagnostic score reporting that will help you identify your strengths and weaknesses so you'll be ready on exam day.

Whether you use this book throughout the school year or as a refresher in the final weeks before the exam, REA's *Crash Course* will show you how to study efficiently and strategically, so you can boost your score.

Good luck on your AP® Biology exam!

ABOUT OUR AUTHORS

Michael D'Alessio earned his B.S. in Biology from Seton Hall University and his M.S. in Biomedical Sciences from the University of Medicine and Dentistry of New Jersey. He has had an extensive career teaching all levels of mathematics and science, including AP® Biology. Currently, Mr. D'Alessio serves as Director of Science, Instructional Technology, and Assessment at Watchung Hills Regional High School in Warren, New Jersey.

Lauren Gross earned her B.S. in Biology from Dickinson College and her Ph.D. in Plant Physiology from Pennsylvania State University. She taught AP® Biology to homeschooled children in the United States and abroad for Pennsylvania Homeschoolers. As an assistant professor at Loyola College in Maryland, Ms. Gross taught various biology, genetics, and botany courses.

Jennifer C. Guercio earned her M.S. in Molecular Biology with a concentration in neuroscience from Montclair State University. For the past several years, she has been doing research in neuroscience as well as teaching academic writing at Montclair State. Ms. Guercio was a Park Scholar at North Carolina State University, where she earned her B.A. and M.A. degrees.

Christina Palffy earned her B.S. in Biology from the University of Illinois in Urbana-Champaign and her M.S. in biology from Northeastern Illinois University. She has had an extensive career teaching botany, general biology, environmental biology, and genetics. Currently, she serves on the board of the Illinois Junior Academy of Science and teaches AP® Biology at Adlai E. Stevenson High School in Lincolnshire, Illinois.

FOREWORD

The AP® Biology examination will be a measure of how much you have learned throughout the year in your AP® Biology class. This *Crash Course* has been written specifically to help you achieve success on the exam. It covers all the material and themes—the "Big Ideas"—that are stressed throughout the current AP® Biology course framework. There is also additional material on the inquiry-based Laboratory Investigations, and pointers on how to write a comprehensive essay for the free-response section.

REA's *AP® Biology Crash Course* will give you an idea of how well prepared you are before taking the exam. You will be able to determine for yourself which concepts will require additional study. Keep in mind this is *not* a textbook, but rather a unique way to approach your preparation for the exam. The material is presented in a convenient outline format and includes numerous illustrations to help you better understand the material.

The AP® Biology examination is a cumulative test based upon a year-long course of study. With this *Crash Course*, you're well on your way to achieving success on the AP® Biology exam.

Best effort,

E. A. Struzziero
AP® Biology Teacher
Lynnfield High School
Lynnfield, Massachusetts

ABOUT REA

Founded in 1959, Research & Education Association (REA) is dedicated to publishing the finest and most effective educational materials—including study guides and test preps—for students of all ages.

Today, REA's wide-ranging catalog is a leading resource for students, teachers, and other professionals. Visit *www.rea.com* to see our complete catalog.

ACKNOWLEDGMENTS

We would like to thank Larry B. Kling, Editorial Director, for his overall guidance; Pam Weston, Publisher, for setting the quality standards for production integrity and managing the publication to completion; John Cording, Technology Director, for coordinating the design and development of the REA Study Center; and Wayne Barr, Test Prep Project Manager, for editorial project management.

We would also like to extend special thanks to Mark Little for technically reviewing the manuscript; Ellen Gong for proofreading; Kathy Caratozzolo of Caragraphics for typesetting this edition; and Jennifer Calhoun for file prep.

PART I

INTRODUCTION

Keys for Success on the AP® Biology Exam

I. Using the *AP® Biology Crash Course* to Prepare for Success

Congratulations on your decision to take AP® Biology. Taking an AP® course, especially an AP® science course, is a challenging endeavor. AP® courses represent college-level classes that you take in high school, so they are supposed to be challenging. But with your hard work, the help of your teacher, and this test prep, you will be guided toward success on this exam and in college.

REA's *Crash Course* is designed with you, the student, in mind. Based on a careful analysis of the most recent College Board AP® Biology Course and Exam Description, the book covers the key topics and concepts you need to know to score well on the exam.

Using this *Crash Course* as a supplement to your coursework, and as a final review in the last few weeks before the exam, will ensure that you are well-prepared for test day.

Now let's look at the exam!

1. Overview

The AP® Biology course is focused on building students' understanding of biological concepts and developing their reasoning skills in a scientific laboratory setting. It is based on 4 Big Ideas that are designed to help you understand core scientific principles and other biological concepts. The course also includes inquiry-based laboratories that are aligned with the learning objectives of the course.

The AP® Biology course is the equivalent of a two-semester college-level introductory biology course. To succeed on the exam, you need to master the key concepts that make up the 4 Big Ideas and apply these concepts to various situations in a traditional test format.

Big Idea 1: **Evolution**—The process of evolution drives the diversity and unity of life.

Big Idea 2: **Energetics**—Biological systems use energy and molecular building blocks to grow, reproduce, and maintain dynamic homeostasis.

Big Idea 3: **Information Storage and Transmission**—Living systems store, retrieve, transmit, and respond to information essential to life processes.

Big Idea 4: **Systems Interactions**—Biological systems interact, and these systems and their interactions exhibit complex properties.

2. The Structure of the Exam

The AP® Biology exam is made up of two sections: multiple-choice and free-response. Each section includes questions that test students' understanding of the 4 Big Ideas.

The exam is 3 hours long. It is comprised of 60 multiple-choice questions and 6 free-response questions.

Section I: Multiple-choice = 50% of the exam grade

You will have 90 minutes to answer the 60 multiple-choice questions.

The multiple-choice questions will consist of four answer options (A through D). Your score will be based upon the number of correct responses you give. No scoring penalties are imposed for incorrect or unanswered questions.

The AP® Biology course content is organized into 8 commonly taught units that provide one possible sequence for teaching the course. These units and their exam weighting for the multiple-choice section of the exam are listed in the following table.

AP® Biology Unit	Exam Weighting
Unit 1: Chemistry of Life	8%–11%
Unit 2: Cell Structure and Function	10%–13%
Unit 3: Cellular Energetics	12%–16%
Unit 4: Cell Communication and Cell Cycle	10%–15%
Unit 5: Heredity	8%–11%
Unit 6: Gene Expression and Regulation	12%–16%
Unit 7: Natural Selection	13%–20%
Unit 8: Ecology	10%–15%

Section II: Free-Response Section = 50% of the exam grade

You will have 90 minutes to answer 4 short free-response questions and 2 long free-response questions.

To achieve a high score on the free-response questions, you must provide ample scientific reasoning, relevant examples, and other appropriate evidence to support their answers. The 2 long questions focus on evaluating and interpreting results from experiments. One of the two requires graphing. The 4 short questions involve scientific investigation, conceptual analysis, model or visual analysis, and data analysis.

AP® Biology Exam Format at-a-Glance

SECTION I		
Question Type	Number of Questions	Timing
Multiple-Choice	60	90 minutes
SECTION II		
Question Type	Number of Questions	Timing
Long Free-Response	2	90 minutes
Short Free-Response	4	

You are permitted to use a four-function calculator (with square root), scientific, or graphing calculator to answer questions on both sections of the exam since both sections contain questions requiring data manipulation. To see which types of calculators are approved for the AP® Biology exam, visit *http://www.collegeboard.org/ap*.

As part of your testing packet, you will be given a list of formulas needed to answer quantitative questions that involve mathematical reasoning.

3. **Scoring the AP® Biology Exam**

   Total scores on the multiple-choice section of the exam are based on the number of questions answered correctly. Points are not deducted for incorrect answers or unanswered questions. The multiple-choice questions are scored by machine, and the free-response questions are scored by AP® exam readers.

   Free-response question scores are weighted and combined with the results of the machine-scored multiple-choice questions to obtain a raw score. This raw score is then converted into a composite AP® score of 5, 4, 3, 2, or 1. These AP® scores determine how qualified students are to receive college credit or placement, as shown in the following table.

AP® Score Qualification
5: Extremely well qualified
4: Well qualified
3: Qualified
2: Possibly qualified
1: No recommendation

4. Using Materials to Supplement Your Crash Course

This *Crash Course* has everything you need to succeed on the exam. However, the College Board's website is also a valuable resource. The site provides information about the test structure, question types, FAQs, and more importantly, additional study materials and sample questions.

5. Some Basic Test-Taking Strategies

One of the best ways to prepare for the AP® Biology exam is to take the online practice exam included with this book. This practice exam will help you become familiar with the format of the test and the types of questions you will be asked. Your instant online score report will provide feedback that will help you to understand which questions give you the most difficulty. Then you can go back to the text of this book, reread the appropriate sections of your textbook, or ask your teacher for help on topics that still give you trouble. The more practice questions you answer in preparation for test day, the better you will do.

You need not study separately for the two sections of the test. As you prepare for the multiple-choice questions, you are also preparing for the free-response questions (FRQs). All of the questions relate back to the topics in the AP® Biology course framework.

On test day, read all the questions carefully, and be alert for qualifying words such as *always, never, not,* and *except.* On the multiple-choice section, review all the answer choices before selecting your answer.

Students ask if they should guess the answer to a multiple-choice question. Remember, there is no penalty for incorrect answers. Therefore, guessing is always advised if you have no idea of the correct answer. Before resorting to a blind guess, however, you should use all your knowledge and understanding of biology to eliminate the possible incorrect answers, so that any guess you are forced to make is an *educated guess.* Of course, you don't have to guess as there are also no points deducted from your score for unanswered questions.

When practicing for the test, give yourself enough time to answer all of the questions. The time left in a given section will be announced by the proctor, but you must use your time wisely. Our online practice test with timed testing conditions will help you budget your time efficiently.

In the free-response section, make sure you write clearly. This sounds simple, but if those who are scoring your exam cannot read your answer, you will not get credit. You should cross out any errors—using a single line through any mistakes—rather than erase them.

When you're in the free-response section, pay particular attention to questions that use the words *justify, explain, calculate, determine, derive,* and *evaluate.* These terms have precise meanings and are there for a purpose. Pay attention and answer the question as it's asked to receive maximum credit. Be sure to support your answer with examples and other scientific evidence. Avoid including irrelevant or extraneous material in your answer.

At this stage of your school career, it may not be necessary to remind you of some basic preparations right before test day—but we will anyway because they're tried-and-true: Get a good night's sleep the night before the exam, eat a good breakfast, and don't forget to bring a bunch of those famous No. 2 pencils.

6. What You Need for Exam Day

Here's a handy chart of what to bring to the exam and what you cannot have in the exam room:

Yes	No
• Several sharpened No. 2 pencils (with erasers) for completing the multiple-choice questions • One or two reliable dark blue or black ink pens for filling out identifying information on the exam booklet covers and for answering the free-response questions. Avoid pens that clump or bleed. • A wristwatch, so you can monitor your time. Make sure it does not beep or have an alarm. • Your school's code if you are testing at a school different from the one you usually attend • Your Social Security number (for identification purposes) • A government-issued or school-issued photo ID and your AP® Student Pack if you do not attend the school where you are taking the exam • The College Board SSD Accommodations Letter if you are taking an exam with approved testing accommodations • Up to two calculators (and extra batteries) with the necessary capabilities for the AP® Biology exam	• Cell phones, smartphones, or any other electronic devices that can access the internet • Cameras or other recording devices • Books, including dictionaries • Scratch paper • Mechanical pencils • Notes you've made in advance • Highlighters and colored pencils • Clothing with subject-related information • Food and drink

PART II

CONTENT REVIEW

UNIT 1
CHEMISTRY OF LIFE

Matter

Key Concepts

A. Because they are composed of matter, the basic rules of chemistry apply to all living organisms.

B. All organisms require an input of energy from the environment, as well as the means to control the orderly use of that energy.

    1. Organisms generally convert the energy they obtain into adenosine triphosphate, or ATP, the cell's "energy currency," which they use to power all life processes, including biosynthesis.

    2. Biochemical reactions, catalyzed by a large array of enzymes that are specific for each reaction, control biosynthesis—the chemical reactions that produce the macromolecules of which an organism's cells are composed.

C. Biological molecules include carbohydrates, lipids, proteins, and nucleic acids that have a number of important functions.

D. Water is the most abundant molecule in living organisms and possesses a variety of instrumental properties that result from its hydrogen bonding.

E. Enzymes are proteins that act as catalysts to speed up biochemical reactions.

II. Water

A. Water has many significant characteristics important to living organisms, primarily because of its polarity and hydrogen bonding. Hydrogen bonding means the attraction between the positive side (hydrogen) of one water molecule and the negative side (oxygen) of another water molecule.

B. Water is a very good solvent of ionic compounds; therefore, many biochemical reactions can take place within a cell and in its immediate environment (the space between cells in a multicellular organism). This is why it is important that cells and their environment are water based.

C. The pH of the aqueous environment inside a cell and its organelles influence many biological activities such as the shape of proteins, the creation of proton gradients across membranes, and the speed at which enzymes catalyze reactions.
   1. Acidic solution—contains more H^+ than OH^- and has a pH below 7.
   2. Basic solution—contains more OH^- than H^+ and has a pH above 7.

D. The following are the properties of water that make it an important molecule, and are due to the hydrogen bonding between water molecules.
   1. *Cohesion*—ability of water molecules to stick together
   2. *Adhesion*—ability of water to adhere to other molecules
   3. *Surface Tension*—difficulty in breaking the surface of water
   4. *High Specific Heat Capacity and Thermal Conductivity*—it heats up and cools down slowly

III. Monomers and Polymers

A. *Monomers* are building blocks of larger macromolecules called polymers.

B. *Macromolecules* are large molecules that fall into four categories: carbohydrates, lipids, proteins, and nucleic acids.

C. *Condensation reactions* are responsible for the biosynthesis of polymers from monomers with the removal of water. The following figure shows the condensation synthesis of a polymer:

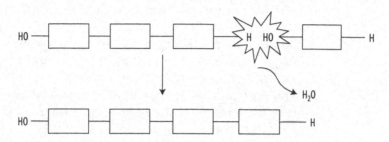

D. Conversely, *hydrolysis reactions* break down polymers into their monomers with the addition of water, and it is the reverse of the reaction shown in the figure on condensation synthesis of a polymer.

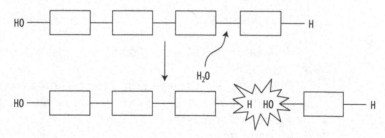

 Biological Molecules

A. These biologically important elements are found in all organisms:

1. *Carbon* (C), *hydrogen* (H), and *oxygen* (O) are found in all macromolecules.
2. Additionally, *nitrogen* (N) is found in significant amounts in proteins and nucleic acids.

3. *Sulfur* (S) is an element commonly found in proteins.
4. *Phosphorous* (P) is prominent in nucleic acids.

B. There are four classes of biological molecules:

1. *Carbohydrates*—consist of sugar (monosaccharides) and polymers of sugars (disaccharides and polysaccharides).
   i. The most important monosaccharide is glucose $(C_6H_{12}O_6)$.
   ii. Sugars are key metabolites used in the synthesis of other organic molecules, as well as the substrates of glycolysis and the products of photosynthesis.
   iii. The particular bonding between carbohydrate subunits is what determines the specific orientation of the carbohydrate and its secondary structure.

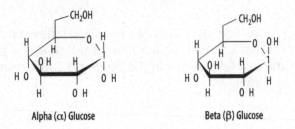

Alpha (α) Glucose   Beta (β) Glucose

2. *Lipids*—water-insoluble molecules composed of glycerol and fatty acids.
   i. *Fats* (triglycerides) are energy storage molecules consisting of one glycerol molecule with three fatty acid molecules attached.
      ➤ *Saturated fatty acids*—do not contain a double bond and are more likely to be a solid at room temperature.

![Structure of a saturated fatty acid]

      ➤ *Unsaturated fatty acids*—have one or more double bonds and are more likely to be fluid at room temperature.

![Structure of an unsaturated fatty acid]

ii. *Phospholipids*—consist of one glycerol molecule with two fatty acid molecules attached as well as a polar component; they can self-assemble into a classic bilayer arrangement that is the basis of all biological membranes.

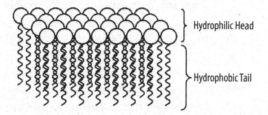

iii. *Wax and steroids*—(including cholesterol) lipids with more complex structures that have a number of functions. The following is the structure of the precursor lipid cholesterol:

3. *Proteins*—polymers made up of different combinations of 20 commonly occurring amino acid monomers.
   i. Proteins have a wide variety of functions, including structural components of cells and tissues, transport proteins in the cell's membranes, and as catalysts called enzymes.
   ii. Amino acids share the same basic structure:

   ➤ Connected by a linear sequence through the formation of peptide bonds by dehydration synthesis

➤ Contain a central carbon atom covalently bonded to four atoms or functional groups:
— One of the four is always a hydrogen atom
— A carboxyl functional group (acidic) –COOH and an amine functional group (basic) –NH$_2$
— Fourth component is a variable R group, which is different for each amino acid

iii. Proteins have four levels of physical structure:
➤ *Primary structure*—refers to the specific linear sequence of amino acids in a polypeptide.
➤ *Secondary structure*—the initial folding patterns of certain lengths of the polypeptide chain, such as alpha helices and beta pleated sheets.
➤ *Tertiary structure*—refers to the overall shape in which a polypeptide eventually folds.
➤ *Quaternary structure*—arises from the association of two or more folded polypeptides to form a multi-subunit protein.

4. *Nucleic acids (DNA and RNA)*—made from monomers called nucleotides.
i. A nucleotide has three parts:

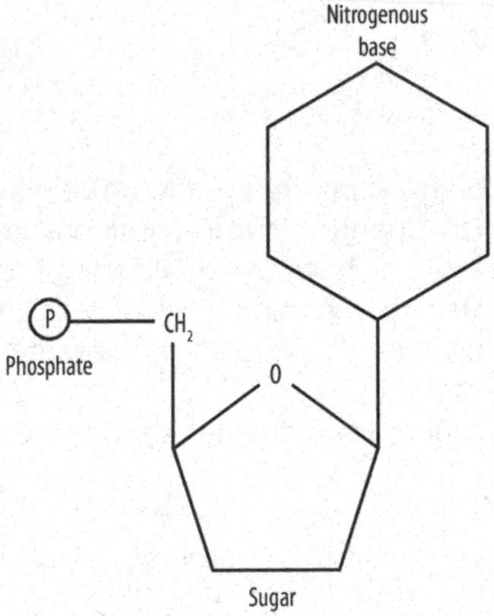

Nitrogenous base

P —— CH$_2$

Phosphate

O

Sugar

➤ A *5-carbon sugar*—either deoxyribose (in DNA) or ribose (in RNA).

➤ A *phosphate group*—present on a nucleotide.

➤ One of the four *nitrogenous bases* is present in each nucleotide.

   — Adenine, thymine, cytosine, and guanine (in DNA)

   — Adenine, uracil, cytosine, and guanine (in RNA)

  ii. Nucleic acids have ends, defined by the 3' and 5' carbons of the nucleotide's sugar. The direction in which other nucleotides are added to the chain during DNA synthesis and the direction in which transcription occurs are both determined by the placement of these two nucleotide ends.

C. Types, Functions, and Examples of Biological Molecules

Type of Biological Molecule	Examples	General Functions
Carbohydrates	Monosaccharides (sugars)	
	Glucose	Energy; building blocks of other carbohydrates
	Deoxyribose and ribose	DNA and RNA
	Polysaccharides	
	Starch and glycogen	Energy storage
	Cellulose	Plant cell wall structure
Lipids	Fats	Energy storage
	Phospholipids	Plasma membrane structure
	Waxes	
	Steroids (cholesterol)	Physical protection
		Hormones (part of cell membranes)

(continued)

Types, Functions, and Examples of Biological Molecules (*continued*)

Type of Biological Molecule	Examples	General Functions
Proteins	Enzymes	Biochemical catalysts
	Other proteins	Structure, movement, signal reception
		Transport of materials in and out of cells
Nucleic Acids	DNA	Storage of genetic information
	RNA	Converts genetic information into proteins
	ATP	Energy currency of the cell

➤ Keep in mind that variations within these biological molecules allow for cells and organisms to possess a much wider variety of functions, such as having different types of hemoglobin or different phospholipids on a cell membrane.

Test Tip

Biologically important molecules are great examples of the "structure and function" theme. Be sure to use these examples in essay questions to gain critical points. However, no specific structures of any monomers or polymers will be asked on the exam.

D. *Enzymes*—proteins that act as catalysts to speed up biochemical reactions.

1. The function of enzymes is to lower the *activation energy* of a reaction. This is the energy required to initiate a chemical reaction.

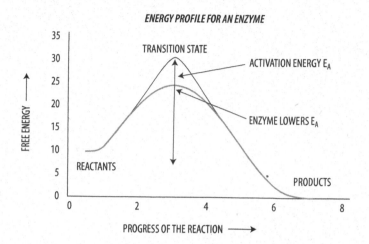

ENERGY PROFILE FOR AN ENZYME

2. The enzyme combines with the substrate or molecule that the enzyme will act upon. The shape of the enzyme's reactive site matches the shape of the substrate molecule.

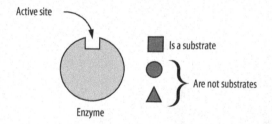

i. When the enzyme and substrate are joined, a catalytic reaction takes place, forming a product. The enzyme can be recycled and used for later reactions.

ii. Enzymes are affected by pH, temperature, and substrate concentration. Enzymes have a pH and temperature optima at which enzyme activity is greatest. Also, as a substrate's concentration increases, the speed at which the reaction occurs increases up to a maximum level at which all enzyme molecules are processing substrate molecules as fast as possible.

iii. *Cofactors* and *coenzymes* can also affect enzyme function. Sometimes the interaction between them causes a structural change, and therefore the enzyme's activity rate changes. Enzymes may also become active only when all necessary coenzymes and cofactors are present. Most coenzymes are vitamins and cofactors are often minerals.

UNIT 2
CELL STRUCTURE AND FUNCTION

Cellular Structure

I. Key Concepts

A. All organisms are made up of one or more cells.

B. The cell is the basic unit of structure and function of organisms.

C. New cells arise only from existing cells by cell reproduction.

D. Cells exchange substances with their environment by transporting these substances in and out of the cell across the plasma membrane.

II. Types of Cells—Prokaryotic and Eukaryotic

A. *Prokaryotic cells* are simpler and more ancient than eukaryotic cells. The following are prokaryotes' important characteristics:

1. No nucleus, only a nucleoid region with one, circular DNA.
2. No membrane-bound organelles.
3. Have a cell wall.
4. Like eukaryotes, prokaryotes contain a plasma membrane, cytoplasm, and ribosomes (location of protein synthesis).
5. No histones or no formation of chromosomes.

B. *Eukaryotic cells* are those of protists, fungi, plants, and animals. They contain the following characteristics:

1. Contain a nucleus, a nuclear envelope to protect DNA, nuclear pores to allow transport into and out of the nucleus, and linear DNA.

2. Have membrane-bound organelles. Plants have chloroplasts, for example, where photosynthesis is carried out, and many plant cells have a large, central vacuole that is absent in most animal cells.

3. Plants have rigid cell walls made of the polymer cellulose, but animal cells do not have cell walls.

4. Like prokaryotes, eukaryotes contain a plasma membrane, cytoplasm, and ribosomes (location of protein synthesis).

5. Contain histones that form into chromosomes.

 III. **Eukaryotic Organelles**

A. Eukaryotic cells maintain internal organelles for specialized functions. Some of these include the following:

Feature	Structure	Function
Mitochondrion (-dria, pl.)	Small organelle with two membranes; inner membranes called *cristae* are folded to increase surface area for electron transport; directly requires oxygen	Site of aerobic respiration; produces ATP; inheritance is always from mother to child
Endoplasmic Reticulum (Rough and Smooth)	Rows of flattened, membranous sacs with or without ribosomes attached Rough ER—has ribosomes Smooth ER—no ribosomes	Sites of protein and membrane synthesis, including detoxification of drugs
Golgi Apparatus	Rows of flattened, membranous sacs	Modifies and transports proteins, etc., for export from the cell; synthesizes carbohydrates
Ribosome(s)	Tiny organelles; no membrane; contain rRNA and protein; bound to ER or float free in cytoplasm	Sites of protein synthesis

(continued)

Feature	Structure	Function
Lysosome(s)	Small, spherical; surrounded by one membrane; contains hydrolytic enzymes	Aids in phagocytosis and intracellular digestion
Vacuoles	Small or large; surrounded by single membrane	Provides turgor pressure for gross plant support; storage of substances
Chloroplast (Type of Plastid)	Various membrane-bound organelles; chloroplast has double membrane plus thylakoids shaped like stacked coins to increase surface area	Site of photosynthesis; other plastids store starch or fats
Cytoskeleton	Network of microfilaments and microtubules throughout the cytoplasm	Controls cell shape; causes movement of chromosomes and organelles within the cell
Vesicle(s)	Small, spherical, numerous; surrounded by one membrane	Move substances from the ER to the Golgi apparatus and from there to the plasma membrane
Nucleus	Large, round; surrounded by nuclear envelope consisting of two membranes studded with pores	Site of chromosome (DNA) storage and RNA synthesis (transcription)
Cell Wall	Rigid; contains cellulose	Provides support and protection of cells

Membranes and Transport

I. Cell Membrane Structure

A. *Plasma Membrane*—separates internal environment from external environment and allows substances to be transported in and out of the cell.

B. *Selective Permeability*—the plasma membrane is selectively permeable, meaning that it allows some substances to pass through it, but not others. It is a direct consequence of the membrane structure, called the *Fluid Mosaic Model*. Essentially, the membrane is a mosaic of proteins that are embedded in or attached to the phospholipids.

1. The lipid portion of the membrane is composed mainly of phospholipids.
   i. *Phospholipids* have a *hydrophobic* (water-fearing) tail and a *hydrophilic* (water-loving) head.
   ii. The cytosol and the fluid outside the cell (the extracellular fluid) are both aqueous (watery) environments.
   iii. Therefore, phospholipids form a *bilayer*, as shown below, because the hydrophilic heads associate with the cytoplasm and the extracellular fluid, while the hydrophobic tails associate with each other.

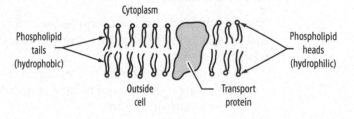

Cytoplasm

Phospholipid tails (hydrophobic)

Phospholipid heads (hydrophilic)

Outside cell

Transport protein

2. *Cholesterol*, also found in membranes, helps regulate membrane fluidity in a way that can be likened to antifreeze in a car: At low temperatures, it keeps membrane lipids from clumping together in a solid-like state, while at high temperatures, it raises the melting point and thus decreases membrane fluidity.

3. *Embedded proteins* in the cell membrane can be hydrophilic, with charged and polar side groups, or hydrophobic, with nonpolar side groups.
   i. The two types of proteins found in membranes are:
      ➤ *Integral Proteins*—transmembrane proteins with hydrophobic and hydrophilic portions
      ➤ *Peripheral Proteins*—bind to integral proteins on the outside of the cell membrane
   ii. Functions of membrane proteins include:
      ➤ transport
      ➤ enzymatic activity
      ➤ signal transduction and cell communication

II. Cell Membrane Transport

A. *Cell Size* is limited by the surface area-to-volume ratio. Cells need a high surface area-to-volume ratio to efficiently exchange materials with the environment.

1. As cells increase in size (volume) the surface area-to-volume ratio decreases as the demand for materials to be imported and exported from the cell increases.

2. To make exchange more efficient, membrane folds might be present to make more surface area for exchange of materials.

B. There are two main types of *cellular transport*: passive transport and active transport.

1. *Passive transport* does not require the cell to use ATP energy and plays a role in both the import of resources and the export of waste.
   i. In *diffusion*, a substance moves down its concentration gradient from an area of higher concentration to an area of lower concentration.

➤ Substances moved are small, uncharged, and non-polar molecules (e.g., carbon dioxide and oxygen).
➤ Substances move directly across the lipid bilayer.

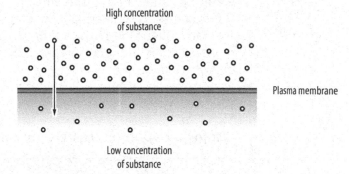

High concentration
of substance

Plasma membrane

Low concentration
of substance

ii. In *facilitated diffusion*, transport proteins move charged molecules (e.g., potassium ions) and larger molecules (e.g., glucose) into and out of the cell.
   ➤ As with diffusion, facilitated diffusion moves a substance down its concentration gradient from an area of higher concentration to an area of lower concentration without the use of ATP.
   ➤ Unlike diffusion, however, the substance moves with the help of carrier proteins or through a channel protein.

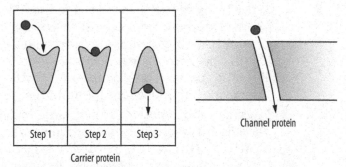

| Step 1 | Step 2 | Step 3 |

Carrier protein

Channel protein

   ➤ Examples of facilitated diffusion include: glucose transport and Na^+/K^+ transport.
iii. *Osmosis* causes water to move across the plasma membrane from a hypotonic solution to a *hypertonic solution*.

➤ *Hypotonic solution* (high water potential, low osmolarity)—has lower concentration of solutes (dissolved substances) than a hypertonic solution (low water potential, high osmolarity).

➤ If two solutions have equal concentrations of solutes, they are called isotonic and there is no net movement of water across the plasma membrane. This is called *dynamic equilibrium.*

➤ In osmosis, the solute molecule is not able to cross the *selectively permeable* plasma membrane.

Osmosis: Hypotonic, Hypertonic, and Isotonic Cells

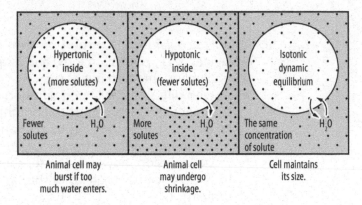

2. *Active transport* requires the cell to use ATP.
   i. Involves movement against the concentration gradient, which is why active transport requires the use of ATP.
   ii. Specific membrane proteins are used in active transport.
   iii. Na^+-K^+ pump is an example of a protein that uses active transport to move ions through the cell's membrane.
3. *Exocytosis* and *endocytosis* move large molecules and food particles across the plasma membrane with the expenditure of ATP; in other words, they utilize active transport.
   i. *Exocytosis*—fusion of vesicles and molecules with the plasma membrane; secretes materials to the outside of the cell.
   ii. *Endocytosis*—the cell takes in the molecules via vesicles that fuse with the plasma membrane.
   iii. *Pinocytosis*—uptake of liquids
   iv. *Phagocytosis*—uptake of solids

UNIT 3

CELLULAR ENERGETICS

Energy

I. Chemistry of Life—Energy Changes

A. *Energy*—capacity to do work.

1. *First Law of Thermodynamics*—energy can neither be created nor destroyed but can change from one form to another and be transferred.

   Example: Plants convert light energy from the sun to make glucose, a form of chemical energy.

2. *Second Law of Thermodynamics*—every energy transfer increases entropy of the universe (disorder).

   i. All living systems will not violate the Second Law of Thermodynamics.

B. *Free Energy*—energy available in a system to do work; organisms need this free energy to maintain organization, to grow, and to reproduce.

1. *Exergonic* reactions release free energy.

$$AB \rightarrow A + B + Energy$$

   i. In *catabolic reactions*, reactant(s) are broken down to produce product(s) containing less energy.
   ii. The energy released can be used for reactions that require energy.

2. *Endergonic* reactions require free energy.

$$A + B + Energy \rightarrow AB$$

   i. In *anabolic reactions*, reactant(s) are joined to produce product(s) containing more energy.

      ii. The free energy required by anabolic reactions is often provided by ATP produced in catabolic reactions.

3. *Adenosine triphosphate (ATP)* carries energy in its high-energy phosphate bonds.

      i. ATP is formed from adenosine diphosphate (ADP) and inorganic phosphate.

$$ADP + Pi + Energy \rightarrow ATP$$

      ii. Conversely, when ATP is broken down into ADP and Pi via hydrolysis, energy is released (*exergonic*) that can be used in *endergonic* reactions.

      iii. In addition, ATP can donate one of its phosphate groups to a molecule, such as a substrate or a protein, to energize it or cause it to change its shape.

4. Living systems require a consistent input of free energy and an ordered system.

      i. This free energy input allows for a system's order to be maintained.

      ii. If either order in the system or free energy flow ceased to occur, death could result.

      iii. Biological processes are in place to help offset increased disorder and entropy and to help maintain order within a system; therefore, energy input into the system must exceed the loss of free energy to maintain order and to power cellular processes.

      iv. Energy storage and growth can result from excess acquired free energy beyond the required energy necessary for maintenance and order within a system.

      v. Changes in free energy can affect population size and disrupt an ecosystem.

5. *Metabolism*—the totality of all chemical reactions that occur within an organism.

      i. Reproduction and rearing of offspring require free energy beyond what is normally required for the maintenance and growth of the organism. Energy availability can vary, and different organisms utilize a variety of reproductive strategies as a consequence. Some examples include seasonal reproduction by animals and plants and life-history strategy (biennial plants, reproductive diapause).

   ii. Organisms use free energy to help regulate body temperature and metabolism. Mechanisms through which this regulation takes place include:
- *Endothermy*—use of internal thermal energy that is generated by metabolism to maintain an organism's body temperature.
- *Ectothermy*—use of external thermal energy to assist in the regulation of an organism's body temperature.
- Some plant species utilize elevated floral temperatures.

   iii. There is an important relationship between the metabolic rate/unit body mass and the size of multicellular organisms. In other words, smaller organisms generally have higher metabolic rates.

C. Energy Coupling

1. *Coupled reactions*—a chemical reaction having a common intermediate in which energy is transferred from one reaction to another.
2. A system can maintain order by utilizing coupling cellular processes that increase entropy (causing negative changes in free energy) with those that decrease entropy (causing positive changes in free energy).
3. The molecule that is essential for coupling reactions and cellular work is ATP.
4. Exergonic reactions, like ATP $\rightarrow$ ADP, is an example of an energetically favorable reaction because it allows for a negative change in free energy that will then be used to maintain or to increase order within a system that is coupled by reactions that demonstrate changes in positive free energy.
5. The processes of cellular respiration and photosynthesis are coupled to each other. The products of one reaction end up being the reactants in the other.

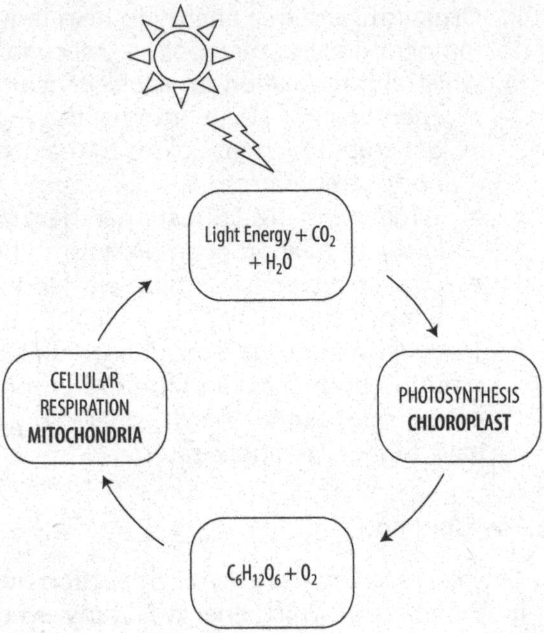
Light Energy + CO_2 + H_2O
PHOTOSYNTHESIS
CHLOROPLAST
$C_6H_{12}O_6 + O_2$
CELLULAR
RESPIRATION
MITOCHONDRIA

6. Electron transport and oxidative phosphorylation are examples of coupled reactions.

D. Modes of Energy Capture

1. Organisms can capture and store free energy for nutritional use in their biological systems.
   i. *Autotrophs*—"self-feeders," create their own organic molecules or food; they are known as producers.
   ii. *Heterotrophs*—cannot create their own organic molecules or food; they are known as consumers.
      ➤ *Hydrolysis*—helps them metabolize carbohydrates, proteins, and lipids as sources of free energy.

➤ The following chart shows modes of nutrition:

Mode of Nutrition	Description; Examples (Other Nonprokaryote Examples)
Photoautotrophy	Use light as an energy source and gain carbon from CO_2; cyanobacteria (also plants and some protists)
Chemoautotrophy	Use an inorganic energy source and gain carbon from CO_2; some archaebacteria
Photoheterotrophy	Use light as an energy source and gain carbon from organic sources; some prokaryotes
Chemoheterotrophy	Use an organic energy source and gain carbon from organic sources; most prokaryotes (also animals, fungi, and some protists)

2. Biological systems can capture energy at multiple points in their energy-related pathways. Some examples of these pathways include the *Krebs Cycle*, *glycolysis*, the *Calvin Cycle*, and *fermentation*.

3. Energy capturing processes, such as NADP+ in photosynthesis and oxygen in cellular respiration, use different types of electron acceptors.

Note: For more on photosynthesis, see Chapter 6; for more on cellular respiration, see Chapter 7.

You will not be required to know the names of enzymes or specific steps and intermediates of pathways. However, be prepared to apply the concepts in this chapter, with particular stress on how organisms utilize free energy.

Photosynthesis

I. Key Concepts

A. Photosynthesis occurs in all photosynthetic autotrophs, including plants, algae, and photosynthetic prokaryotes.

B. In eukaryotes, photosynthesis occurs in chloroplasts; in prokaryotes, it occurs in the plasma membrane and in the cytoplasm.

C. The overall equation for photosynthesis is:

$$6CO_2 + 6H_2O + \text{light energy} \rightarrow C_6H_{12}O_6 + 6O_2$$

D. Photosynthesis is affected by a variety of environmental factors.

E. In eukaryotes, each phase of photosynthesis takes place in the chloroplasts.

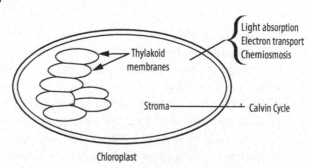

Chloroplast

II. The Two Steps of Photosynthesis

A. Photosynthesis has two main steps:

1. *Light-dependent reactions*—the absorption of light energy and its conversion to the chemical energy of ATP and the reducing power of NADPH.
2. *Light-independent reactions*—the use of ATP and NADPH to convert CO_2 to sugars using the Calvin Cycle.

B. STEP 1—*Light-dependent reactions* occur in the *thylakoid* membranes of chloroplasts in eukaryotes.

1. Pigment molecules collect light energy.
    i. *Chlorophyll a*—main photosynthetic pigment.
    ii. *Chlorophyll b* and *carotenoids*—accessory pigments that allow leaves to capture a wider spectrum of visible light than chlorophyll alone.
    iii. The following graph shows the absorption spectra of photosynthetic pigments:

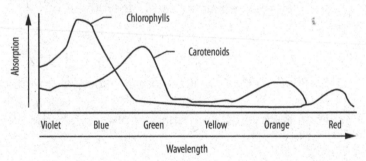

2. *Photosystems (PS) I and II* are embedded in the internal membranes of chloroplasts (thylakoids) and consist of hundreds of pigment molecules that funnel light energy to two *chlorophyll a* molecules at the reaction center of each photosystem. Essentially, they use an electron transport system to transfer higher free energy electrons.
3. *Electron Transport Chain (ETC)*—an electrochemical gradient of hydrogen ions (protons) across the thylakoid membranes that undergoes redox reactions in a series.

    i. Electrons are transferred from PSII → primary electron acceptor → until donated to PSI → next electron carriers → donated to NADP⁺ to reduce it to NADPH.

    ii. Electrons can take either a noncyclical route or a cyclical one. The primary difference between the two is that the cyclical flow of electrons produces more ATP and takes place because the Calvin Cycle uses more ATP per mole than NADPH per mole, and hence replenishes the used ATP.

    iii. The proton gradient is linked to the synthesis of ATP and ADP and inorganic phosphate via ATP synthase.

4. *Chemiosmosis*—the movement of H^+ ions down their concentration gradient from inside the thylakoids to the stroma. As they do this, they pass through the enzyme, *ATP synthase*, which causes the catalysis of ATP from ADP and Pi.

    i. The following figure shows both the electron transport and chemiosmosis of photosynthesis.

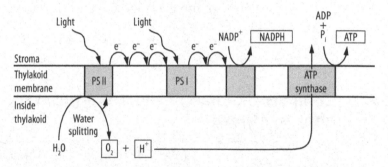

5. *Flow summary*—absorption and conversion of light energy to ATP and NADPH.

    i. Pigments absorb light energy.

    ii. Light energy sends electrons down the electron transport chain.

    iii. The electrons eventually reduce NADP⁺ to NADPH.

    iv. Water is split, forming e⁻, H⁺, and O_2.

    v. H⁺ concentration builds up inside the thylakoids (the thylakoids space).

    vi. When H⁺ move through ATP synthase from the thylakoid space to the stroma, ATP is formed.

    vii. NADPH and ATP are used in the second step of photosynthesis, carbon fixation.

C. STEP 2—*Calvin Cycle/Light Independent Cycle*

1. Occurs in the stroma of chloroplasts.
2. Uses the products (ATP and NADPH) to produce glucose
3. The following figure depicts the steps in the *Calvin Cycle*:

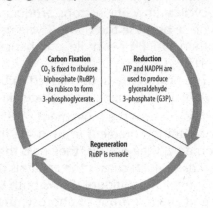

i. Six turns of the cycle fix six carbons, representing one molecule of glucose.
ii. Six turns of the cycle require 18 ATP and 12 NADPH.

 III.

Comparison Chart: Cellular Respiration and Photosynthesis

Process	Takes Place in Cellular Respiration	Takes Place in Photosynthesis
Breakdown of glucose	' Yes	No
Synthesis of glucose	No	Yes—Calvin Cycle
O_2 is released	No	Yes—light-dependent reaction
O_2 is consumed	Yes—ETC and oxidative phosphorylation	No

(continued)

Comparison Chart: Cellular Respiration and Photosynthesis (*continued*)

Process	Takes Place in Cellular Respiration	Takes Place in Photosynthesis
Chemiosmosis	Yes—ETC	Yes—ETC
CO_2 is released	Yes—shuttle step and Krebs Cycle	No
CO_2 is consumed/ fixed	No	Yes—Calvin Cycle
ATP is produced	Yes—glycolysis, Krebs Cycle, ETC, and oxidative phosphorylation	Yes—light-dependent reaction
ATP is consumed	Yes—glycolysis initial investment	Yes—Calvin Cycle
Pyruvate as intermediate	Yes—glycolysis	No
NADH produced	Yes—glycolysis, shuttle step, Krebs Cycle	No
NADPH produced	No	Yes—light-dependent reaction

Test Tip

Familiarizing yourself with the similarities between cellular respiration and photosynthesis is recommended. It's been a popular test item on past AP® Biology exams. Be sure you know about ATP production, electron transport use, compartmentalization between chloroplast and mitochondria, hydrogen and electron acceptor molecules, such as NADH, $FADH_2$, and NADPH.

Fermentation and Cellular Respiration

Key Concepts

A. Cellular respiration is the catalysis (breakdown) of glucose to produce energy (ATP) and organic intermediates used in the synthesis of the other organic molecules (amino acids, lipids, etc.) needed by the cell.

B. Some form of cellular respiration takes place in nearly all organisms.

   1. *Glycolysis* is the oldest metabolic pathway, is virtually universal, and takes place in the cytoplasm of cells.
   2. Aerobic respiration—the Krebs (or citric acid) Cycle, electron transport, and chemiosmosis—takes place in mitochondria in eukaryotes.

C. Refer to this overall equation for cell respiration:

$$C_6H_{12}O_6 + 6O_2 \rightarrow 6CO_2 + 6H_2O + Energy$$

Although this equation is almost the reverse of the equation for photosynthesis, the two processes involve different enzymes and biochemical pathways, as well as different organelles.

D. Cells may utilize an anaerobic pathway (fermentation) that does not require O_2, or an aerobic pathway that does require O_2.

   1. Glycolysis is the first step of both pathways. This step does not require O_2.

2. Aerobic respiration has three additional steps, the second of which requires O_2, as the final electron receptor of the electron transport chain.

    i.   The Krebs Cycle (citric acid cycle) takes place in the matrix of the mitochondria.

    ii. The electron transport chain takes place in the inner membrane of the mitochondria.

    iii. Chemiosmosis takes place across the inner membrane of the mitochondria.

    iv. The following figure shows the location of fermentation and the steps of cellular respiration in the mitochondrion:

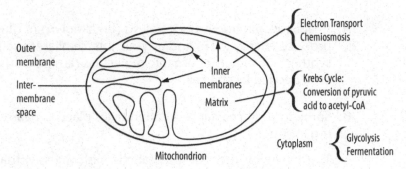

Mitochondrion

3. Anaerobic fermentation has one additional step following glycolysis that regenerates the oxidizing agent, NAD^+, to allow glycolysis to operate in the absence of O_2.

4. Both fermentation and cellular respiration are catabolic and involve oxidation-reduction reactions:

    i.   Loss of electrons is oxidation (glucose to carbon dioxide).

    ii. Gain of electrons is reduction (oxygen to water).

    iii. Electrons = Energy.

Test Tip

Anaerobic fermentation produces much less energy than aerobic respiration: only 2 ATP per glucose processed, as compared to 36 ATP produced by aerobic respiration per glucose molecule.

II. The Four Parts of Cellular Respiration

A. *Glycolysis*—a ten-step metabolic pathway, catalyzed by a series of enzymes, which breaks one glucose molecule down to two molecules of pyruvic acid (pyruvate).

   1. The following are the most important features to remember about the glycolysis reaction series:

      i. The bonds of the glucose molecule are rearranged.

      ii. NAD^+ is reduced to NADH, one of the two electron carriers in cellular respiration.

      iii. Free energy is then released in the form of ATP—which comes from ADP and inorganic phosphates.

      iv. Pyruvic acid is produced and is transported from the cytoplasm to the mitochondria for future oxidation.

      v. The following figure shows the process of glycolysis and its end products:

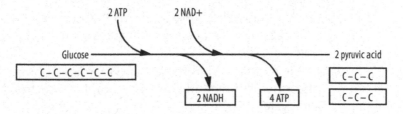

B. *Shuttle Step*—the conversion of pyruvic acid to acetyl-CoA occurs in the matrix of the mitochondria and involves three important features:

   1. Coenzyme A (CoA) is added.

   2. Pyruvate is oxidized, producing NADH.

   3. The 3-carbon pyruvate is converted to the 2-carbon acetyl-CoA, releasing a molecule of CO_2.

4. The following figure shows the conversion of pyruvic acid to acetyl-CoA:

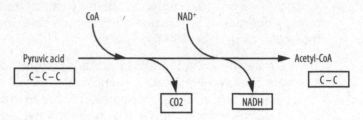

C. *Krebs (or citric acid) Cycle*—series of reactions that continually regenerates one of its first reactants, *oxaloacetic acid* and contains the following important features:

1. The Krebs Cycle (citric acid cycle) produces the majority of NADH, $FADH_2$, and CO_2 (waste product) for cellular respiration.
2. NADH and $FADH_2$ carry electrons that were extracted during the Krebs Cycle reactions and carries them to the electron transport chain.
3. ATP is synthesized from ADP and inorganic phosphates via phosphorylation, and coenzymes capture electrons during this cycle.
4. The key intermediate, *oxaloacetate (OAA)* is added to acetyl-CoA to make *citrate*, which starts the entire Krebs Cycle.

D. *Electron Transport Chain*—carried out by electron carriers that undergo a series of redox reactions as electrons are passed from one carrier to another.

1. The following are the important features to know about electron transport:
   i. Occurs in the mitochondria and is similar to the ETC of photosynthesis.
   ii. NADH and $FADH_2$ deliver electrons and pass them to a series of electron acceptors as these electrons move toward the final electron acceptor, O_2.
   iii. This passage of electrons within the mitochondria is accompanied by a proton gradient that facilitates the electrons' movement down the ETC.

iv. In the ETC, no direct ATP is made; it must be coupled to *oxidative phosphorylation* via chemiosmosis (or the diffusion of H^+ ions across the membrane).

➤ Cellular respiration produces a total of 38 ATP.

III. Fermentation

A. Glycolysis is the first step of both aerobic respiration and fermentation.

1. Two of the end products of glycolysis, pyruvic acid and NADH, can be processed anaerobically in the cytoplasm of certain cells.
2. The second step of fermentation does not produce ATP directly.
3. Rather, it generates NAD^+, which is required to keep glycolysis running and producing ATP.

B. Lactic acid fermentation includes glycolysis plus an additional reaction that generates NAD^+ and lactic acid.

1. Certain fungi, bacteria, and muscle cells have special enzymes that carry out lactic acid fermentation.
2. In vigorously exercising muscle cells, lactic acid fermentation provides ATP when the circulatory system cannot keep up with the oxygen demands of the muscle cells.

C. Alcohol fermentation includes glycolysis plus additional reactions that produce NAD^+, ethanol, and CO_2.

1. Single-celled organisms, such as yeast and some plant cells, have special enzymes to carry out alcohol fermentation.
2. Yeast is used in bread making because CO_2 gas causes bread to rise; the ethanol is removed by subsequent baking. Yeast is also used in beer making because it produces ethanol; CO_2 in an enclosed container produces carbonation.

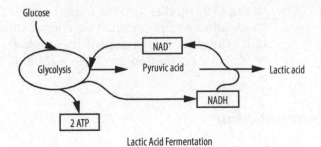

Lactic Acid Fermentation

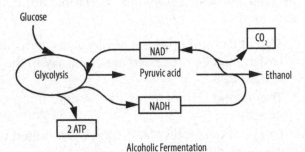

Alcoholic Fermentation

There is usually at least one AP® Biology exam question that requires comparing photosynthesis and cellular respiration. Also, keep in mind that plants carry out both photosynthesis and cellular respiration. Be sure to review the photosynthesis and cellular respiration chart found in Chapter 6: Photosynthesis. Conceptually understand the inputs, outputs, and purpose for major steps in both cell respiration and photosynthesis.

UNIT 4
CELL COMMUNICATION
AND THE CELL CYCLE

Homeostasis

I. System Equilibrium

A. In all organisms, there are mechanisms to achieve *homeostasis*, the maintenance of stable internal conditions in a system, such as a cell, organism, or ecosystem.

B. Many aspects of the environment can change, and organisms must respond accordingly.

1. *Gas levels*—Oxygen is constantly being used in cellular respiration and carbon dioxide is being produced. Cells must have a way of gaining oxygen while excreting carbon dioxide.
   i. Without enough oxygen, ATP cannot be produced.
   ii. With too much carbon dioxide, the pH of the cell will get too low.

2. *Osmolarity*—The balance of water and solutes must be maintained due to osmosis.
   i. If extracellular fluids are hypotonic (low in solute concentration) compared to the cell, water will enter the cell and may cause it to burst (lysis).
   ii. If extracellular fluids are hypertonic (high in solute concentration) compared to the cell, water will leave the cell and perhaps cause it to dry out (crenation).

3. *Temperature*—Cellular structures and the rates of crucial reactions can be affected by changes in temperature.
   i. If temperatures are too low, cellular reactions may happen at too low of a rate to be effective.
   ii. If temperatures are too high, chemical structures, like the tertiary structures of proteins, get disrupted and become nonfunctional.

C. There are three structures necessary to respond to any homeostatic disruption.
   1. The system needs a way to detect the change in the internal or external environment. This is called a **sensor** (or receptor).
   2. The system needs a way to initiate a response to the stimulus. This has to be a cell tissue or organ that is the **integrator**.
   3. A response to the stimulus has to occur. This happens by an **effector** that is an organ, cell, or tissue that gets information from the integrating center and causes a change to help maintain homeostasis.

D. In multicellular organisms, cells have to work together in order for homeostasis to be maintained. This can be done through hormones being transported from one part of the organism to another, or, in animals, a combination of hormones and communication through the nervous system.
   1. An example of maintaining homeostasis in plants involves water conservation.
      i. When soil is dry, cells in the roots act as a sensor.
      ii. Plant root cells, acting as the integrator, will release a hormone called abscisic acid that travels up to leaves.
      iii. This causes *guard cells* to close water-losing holes called *stomata* to conserve water. The action of the root cells is to release abscisic acid, and the response is the closing of the stomata, which are the effectors.
   2. An example of maintaining homeostasis in humans is thermoregulation, or maintaining body temperature.
      i. The sensors within the brain's hypothalamus detect changes in body temperature.
      ii. Integrators in the brain will trigger a response that will cause shivering muscles, which will produce heat if the temperature is cold. If the body is hot, then one response will be to stimulate sweat glands to cool it by evaporative cooling. The sensor and integration center of the hypothalamus controls this response.
      iii. The effectors are either the muscles (when cold) or sweat glands (when hot).

E. Organisms use both negative and positive feedback mechanisms to maintain their internal environments, respond to external stimuli, and regulate growth and reproduction.
   1. *Negative feedback*—occurs when a stimulus produces a result and the result inhibits further stimulation.
      i. Thermoregulation is an example of negative feedback.
      ii. The body gets colder and the response is to warm the body.
      iii. Once the body is at the homeostatic temperature, the response will stop.
   2. *Positive feedback*—occurs when a stimulus produces a result, and the result causes further stimulation, thereby triggering an event.
      i. An example is labor and childbirth.
      ii. When a baby's head pushes on the uterine walls (the stimulus), a hormone called oxytocin is released which causes muscles to contract which causes *more* pressure on the uterine wall.
      iii. This occurs until the stimulus is gone after the baby exits the body.

F. Homeostatic mechanisms that are conserved or divergent across related organisms reflect either continuity due to common ancestry or evolutionary change in response to distinct selective pressures.

II. Disruptions of Homeostasis

A. Biological systems are affected by disruptions in their dynamic homeostasis. These disruptions can occur at the molecular level and affect the health of the organism, or at the level of the entire ecosystem, affecting the survival of parts or all of the population.

B. Molecular and Cellular Disruptions of Homeostasis
   1. Foreign toxic substances enter the organism and are detected by multiple cells, tissues, and organs in order to expel them from the organism and reinstate homeostasis.

2. The body combats these disruptions through physiological and immunological responses.

3. When homeostasis cannot be maintained in a human or animal, medical interventions are needed. If that is not possible, disease and/or death will occur.

C. Disruptions to an Ecosystem's Homeostasis—these disruptions can come from the following sources:

1. *Other species*—can disrupt the homeostasis of another species through predation or parasitism. An example is the introduction of the Emerald Ash Borer in North America. North American trees had no defenses against this Asian beetle, and it has decimated forests and decreased the biodiversity once maintained in the forests.

2. *Direct human impact*—humans can impact not only their ecosystems but also those of other species. An example is contamination of a local lake that is not only a water source for humans but also home to fish, insects, and birds that are also affected by the contamination.

3. *Adverse natural events*—such as hurricanes, earthquakes, or floods can change the environment greatly and destroy habitat and organisms changing the homeostatic balance. An example is a hurricane that floods an island and destroys the land-dwelling insects in that area. Not only is living space for terrestrial animals gone, but food sources can be gone as well.

4. *Limitation of water*—All organisms need water. Those that have evolved in dry areas have mechanisms that help them to survive despite the limitation. If water becomes limited suddenly, organisms do not have mechanisms allowing them to maintain homeostasis in its absence.

Cell Cycle and Mitosis

I. Key Concepts

A. The number of chromosomes an organism has in its body cells does not vary from cell to cell, or organism to organism of the same species.

B. The genetic material (DNA) of all organisms is contained in chromosomes that become especially compact during cell reproduction.

C. The cell cycle is the life cycle of a cell and includes periods when a cell is not dividing as well as those when it undergoes cell division.

D. *Sister chromatids* consist of two duplicated chromosomes held together at the centromere.

E. *Mitosis* occurs in eukaryotes and produces cells with nearly identical genetic makeup.

   1. Mitosis is used for the purpose of organismal reproduction in single-celled organisms.
   2. It is used for purposes of development and cell replacement in the normal growth and maintenance of the bodies of multicellular organisms.

F. Prokaryotes generally reproduce by a process called *binary fission.*

II. The Cell Cycle

A. The two main phases of the cell cycle are interphase and cell division.

B. *Interphase*—Cells spend most of their time in interphase.

1. *G_1 phase*—The first phase of interphase is G_1 during which the new cell grows to mature size and may begin to carry out its specific function.
2. *S phase*—If the cell is going to divide again, it duplicates its chromosomes during the S phase by the process of DNA replication.
3. *G_2 phase*—Once the DNA is replicated, the cell enters G_2 during which it prepares for cell division.
4. *G_0 phase*—Some cells do not divide, or they delay division; these cells enter the G_0 phase sometime during G_1.

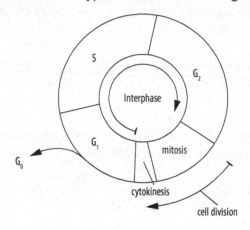

C. The second main phase of the cell cycle is cell division, which includes mitosis and cytokinesis.

III. Mitosis

A. *Mitosis*—the division of the nucleus, which leads to the separation of the chromosomes that were previously duplicated in the S phase to produce two chromatids that are attached at their centromeres; it has four main stages.

B. Phases of Mitosis

1. Prophase—
   i. The chromatids condense.
   ii. The nuclear membrane breaks down and disappears.
   iii. A cytoskeletal structure called the *mitotic spindle* forms and is used to pull the chromatids apart to either pole of the cell.
      ➤ Two *centrosomes* are synthesized.
      ➤ In animal cells, the centrosomes contain small cylindrical bodies called *centrioles*.
      ➤ The centrosomes move to opposite sides of the cell.
      ➤ Spindle fibers, made of microtubules, radiate outward from the centrosomes and some, called *kinetochore fibers*, attach to each chromosome at its centromere.

2. *Metaphase*—During metaphase, the chromatids line up across the center of the cell which is called the *metaphase plate*.

3. *Anaphase*—Chromatids separate from one another during anaphase, at which point each chromatid is now considered to be an individual chromosome.
   i. The centromeres of the chromatids split.
   ii. The kinetochore fibers pull one copy of each chromosome to one pole and the rest to the other side of the cell.

4. *Telophase*—the last stage of mitosis.
   i. The mitotic spindle disassembles.
   ii. The chromosomes unwind from their highly compacted state.
   iii. A new nuclear membrane forms and surrounds each new complete set of chromosomes.

  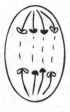 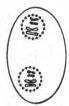

| Prophase | Metaphase | Anaphase | Telophase |

IV. **Cytokinesis**

   A. *Cytokinesis*—the division of the cytoplasm following mitosis whereby the two newly formed nuclei become incorporated into separate cells.

      1. In animal cells, a special collection of microfilaments of the cytoskeleton form a *cleavage furrow* in the center of the cell, which causes the cell to be pinched into two cells.

      2. In plant cells, vesicles from the Golgi apparatus form a *cell plate* in the center of the cell along which new cell wall material is deposited between the two newly forming plasma membranes.

Cytokinesis in Plant and Animal Cells

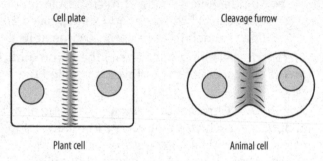

Plant cell   Animal cell

V. **Regulation of the Cell Cycle**

   A. The cell cycle is regulated to help prevent the production of abnormal cells that could eventually become cancerous.

   B. Some Ways That the Cell Cycle Is Regulated

      1. *Checkpoints*—essential points during the cell cycle that regulate the process of passing from one stage to the next.

      2. G_0 *phase*—a nondividing stage of the cell cycle that halts the cycle from proceeding.

      3. *Growth factor*—protein/hormone that promotes the division of cells.

4. *Density-dependent inhibition*—process in which cells stop dividing when they are in contact with each other.
5. *Anchorage dependence*—cells must be attached to something to divide properly.

C. When Regulation of the Cell Cycle Does Not Work

1. *Cancer cell*—cells that are said to be "transformed" from normal cells to cancer cells and do not exhibit density-dependent inhibition. Have uncontrolled growth pattern.
2. *Tumor*—a pocket of abnormal cells among normal cells.
3. *Benign tumor*—nonspreading of abnormal cells.
4. *Malignant tumor*—abnormal cells that invade and impact the normal function of an organ.
5. *Metastasis*—spreading of malignant tumor to other parts of the body.

Cell Communication

I. Key Concepts

A. Cell-to-cell communication is essential for multicellular organisms and their overall development, growth, and homeostasis.

B. Cell-to-cell communication between cells is also important for unicellular organisms.

C. Universal mechanisms of cell communication suggest an evolutionary similarity among species.

II. Evolutionary Similarities

A. Homeostasis—the maintenance of stable internal conditions in the body—is generally controlled by a group of body sensors, the nervous system, and the endocrine system that control in unison several body systems (e.g., thermoregulation involves the hypothalamus sensing body temperature, causing shivering of muscles and constriction of blood vessels to the skin, which produces sweat if the body is hot). For more, see Chapter 8: Homeostasis.

　1. Organisms use feedback mechanisms through cell communication to respond to changes in the internal or external environment.

　2. Negative feedback involves a signal pathway that ends up with a response that counteracts the stimulus or change.

3. Positive feedback involves a response that amplifies the initial stimulus or change.

B. Cells can communicate with each other in some of the following ways:

1. *Chemical messengers*—such as hormones
2. *Cell-to-cell contact*
3. *Synaptic signaling*—neurotransmitters diffuse across a synapse to a single cell

C. The three stages of cell signaling are *reception, transduction, and response.*

1. *Reception*—chemical signals bind to cellular protein.
2. *Transduction*—binding leads to a change along a signal transduction pathway.
3. *Response*—a specific cellular activity is triggered.

D. *Signal Transduction Pathway*—the process by which a signal on the cell's surface is converted to a specific cellular response—is strikingly similar in yeast and animal cells. Their evolutionary similarities in cell communication still exist today, despite the fact that the common ancestor of yeast and animals lived more than one billion years ago. Even signaling between bacteria and plants is similar in some ways.

III. Local and Long-Distance Signals

A. Communication between cells involving *chemical messengers* can be *local* or *long distance.*

1. *Local regulators* are secreted by cells and only affect the activity of nearby cells.
   i. Examples include *neurotransmitters* in the nervous system, *histamine* in the immune system, *growth factors* in development, and *prostaglandins* in the immune and reproductive systems.

    ii. Local regulators act quickly and do not enter the bloodstream.

  2. *Hormones* act over long distances because they enter the circulatory system and are transported around the body. Examples include insulin, testosterone, and estrogen, which are all regulated by the endocrine system.

B. When a hormone comes in contact with a *target cell*, it either enters the cell or binds to receptors on the surface of the cell.

  1. Most chemical messengers, including most *protein-based hormones*, bind to *proteins* that act as *receptors* on the *plasma membrane* of target cells.

    i. The binding between a hormone and a receptor causes a physical change (usually involving movement) of the protein receptor, which sets in motion a message relay system inside the cell called a *signal transduction pathway*.

    ii. A series of *secondary messengers* carry the signal until it eventually results in a response by the cell.

    iii. Common signal transduction pathways include *protein modifications* such as how methylation changes the signaling process, *protein phosphorylation*, activation of *G-proteins* and *cyclic AMP*, or increases in *calcium ion* levels.

    iv. Changes in signal transduction pathways can alter cellular response, and in some cases where the pathway is blocked or defective, the changes can become deleterious, preventative, or even prophylactic.

      ➤ An example would include an organism being bit by a poisonous spider that injects a type of toxin inside the organism's body; the poison blocks either a specific transduction pathway or a series of them. Effective treatment of the bite includes removal of the block to restore the health of the organism.

      ➤ Sometimes creating a block of a signal transduction pathway is the goal, so medication like anesthetics is used on the organism to block pain, for example.

  2. The *fat-soluble steroids* and *thyroid hormones* pass through the plasma membrane and bind to receptor proteins within the cytoplasm or the nucleus of cells.

    i.   Hormone binding to a *cytoplasmic receptor* may trigger the response in the target cell.

    ii.   Some steroids *enter* the *nucleus* bound to DNA regulatory proteins and *stimulate transcription* of specific genes.

C. Cells can also *communicate cell to cell*. Examples include antigen-presenting helper T cells, Killer T cells, and plasmodesmata between plant cells that allow the transport of materials from cell to cell.

D. Signal transmission between cells can also affect gene expression. For example, *cytokines* can regulate gene expression by regulating cell replication and division, and *ethylene levels* signal changes in specific enzymes in fruit that indicate that it is time to ripen.

E. Communication between cells also affects cell function. For example, changes in the activity of gene *p53* can cause cancer.

UNIT 5

HEREDITY

Meiosis

I. Asexual and Sexual Reproduction

A. *Asexual Reproduction*—a form of reproduction not requiring meiosis or fertilization; only passes a copy of genes to its progeny. It's a type of reproduction in which there is no variation in genetic makeup. Bacteria reproduce via asexual reproduction.

B. *Clone*—an individual that arises from asexual reproduction.

C. *Sexual Reproduction*—a type of reproduction that involves variation because two parents give rise to their progeny.

Major evolutionary advantage because of genetic variation.

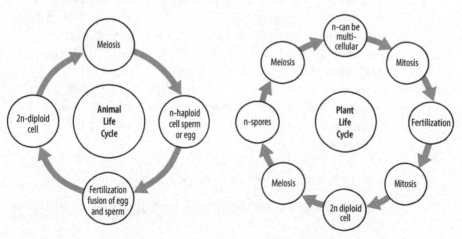

Meiosis

A. Meiosis, like mitosis, is preceded by replication of chromosomes.

B. *Meiosis I* is the first round of cell division, where homologous pairs of chromosomes are separated from one another into two cells that are haploid. Meiosis I has four stages:

1. *Prophase I*—Tetrads are formed or the pairing of homologous chromosomes via *synapsis*; *chiasmata* or the site of crossing over/exchange of genetic material is formed during this phase.
2. *Metaphase I*—homologous chromosomes pair with each other at the metaphase plate.
3. *Anaphase I*—homologous chromosomes separate and sister chromatids stay together.
4. *Telophase I*—the movement of chromosomes to the poles is completed.

C. *Meiosis II* occurs in each of the two new cells after meiosis I is completed. No DNA replication occurs between meiosis I and meiosis II. In meiosis II, the two cells once again go through four stages of division:

1. *Prophase II*—the chromosomes are already condensed and new spindle fibers form.
2. *Metaphase II*—each pair of chromatids lines up in the middle of the cell, and kinetochore fibers attach to the centromeres of each pair.
3. *Anaphase II*—the centromeres holding the chromatids together split, and one chromatid moves to each side of the cell.
4. *Telophase II*—nuclear envelopes re-form around the chromosomes.

D. *Cytokinesis* follows telophase II, resulting in gametes.

1. If sperm are produced, meiosis usually produces four sperm cells.
2. If ova are produced, meiosis often produces a single egg cell, while the other three cells die or have other functions in reproduction.

Every AP® Biology test-taker should know the main differences between meiosis and mitosis. Mitosis produces diploid identical cells that have no genetic variation. Meiosis produces gametes (haploid) that are genetically different because of crossing over in Prophase I of meiosis. Similarly, you should know the stages of mitosis and meiosis and special structures that are formed.

Meiosis I and Meiosis II

Early prophase I

Late prophase I (synapsis begins)

Later prophase I (synapsis has occurred)

Later prophase (cross-over occurring)

Anaphase I
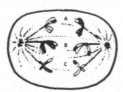

Telophase I
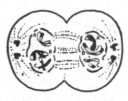

Prophase II
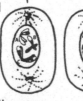

Metaphase II

Anaphase II

After telophase II

III. ## Comparison of Mitosis and Meiosis

Event	Mitosis	Meiosis
DNA Replication	Occurs during interphase	Occurs during interphase
Homologous Chromosomes	Align one after another on metaphase plate	Pair with each other during metaphase I. Align one after another on metaphase plate during metaphase II
Sister Chromatid Separation	Anaphase	Meiosis II Anaphase II
Divisions	1	2
Cells Produced	2 Diploid—genetically identical	4 Haploid—genetically different
Crossing Over	Does not occur	Meiosis I Prophase I

Structure and Inheritance of Chromosomes

I. Key Concepts

A. In the mid-1800s, without knowledge of chromosomes or genes, but with careful experimentation and statistical analysis, Gregor Mendel worked out the basic rules governing how heritable traits are transmitted during sexual reproduction. These rules were Mendel's principles of *heredity*.

B. The *Law of Segregation* is the principle that the two determinants of a characteristic, called *alleles* of a gene, are separated during meiosis and are distributed to separate gametes.

C. The *Law of Independent Assortment* is the principle that the segregation of one set of alleles into gametes, which determine one characteristic, is independent of the segregation of a second set of alleles governing a second characteristic.

D. Genes are located on chromosomes, and in sexually reproducing organisms, alleles—different versions of the same gene—are located at the same position on homologous chromosomes.

E. Understanding the use of probabilities and Punnett squares is important for determining genetic cross outcomes.

F. The sex chromosomes, X and Y, determine the sex of mammals and many insects. Genes on the X and Y chromosomes have special patterns of inheritance called X-linkage and Y-linkage.

G. Genes on the same chromosome do not assort independently if they are linked; linkage can be used to map the relative locations of genes on a chromosome.

H. Pedigree analysis is used to study the inheritance of human genes.

II. Chromosomal Structure

A. Structure and Function of Eukaryotic Chromosomes

Part	Structure	Function
Genes	Made up of the nucleic acid DNA	Will be transcribed onto mRNA Will be translated for proteins
Chromatids	Two replicated chromosomes that are held together at the centromere	Allows proper segregation of chromosome during meiosis and mitosis
Centromere	DNA region found near the middle (not always) chromosome	Hold chromatids together to form a chromosome
Chromatin	DNA and protein combination	Aids in packaging DNA, DNA replication, and expression of proteins
Kinetochore	Proteins	Allows for the attachment of the mitotic spindle to the centromere
Nucleosomes	Histone proteins and DNA	Aids in packaging of DNA
Telomeres	Ends of DNA	Protection against the destruction of the DNA from nucleases

Remember: Eukaryotic DNA is linear, meaning it has definite ends.

Most eukaryotic organisms are diploid. Fungi, such as yeast, can exist as haploid or diploid.

B. Structure and Function of Prokaryotic Chromosome as a Comparison

1. Circular in shape and much smaller than eukaryotic chromosome.
2. Genes are arranged in *operons*—one promoter controlling many genes.
3. Transcription and translation are coupled processes.
4. *Plasmids* are prevalent—extra chromosomal pieces of DNA that carry antibiotic resistance. They are not part of the chromosome. Autonomously replicating.
5. One origin of replication.
6. No histone proteins to condense, but DNA is supercoiled.

III. Inheritance Patterns

A. Terms

1. *Characteristic*—an inheritable feature such as hair color (phenotype).
2. *Trait*—a variant of a characteristic. Example: red or blond hair color.
3. *Allele*—alternative form of a gene, such as tall (*T*) plants are dominant to short plants (*t*).
4. *Dominant allele*—the allele that is fully expressed.
5. *Recessive allele*—the allele that is not expressed.
6. *Genotype*—the genetic makeup of an organism.
7. *Phenotype*—an organism's appearance.

B. *Law of Segregation*—is observed with monohybrid crosses or crosses for a single characteristic. The law states that each trait must result from two distinct factors and that these factors separate from each other during reproduction and are incorporated into separate gametes.

C. *Law of Independent Assortment*—is observed with dihybrid crosses or crossed between two different characters. The law states that alleles assort independently from each other; therefore, dominant alleles can combine with recessive alleles.

D. Genetic Crosses

1. *Monohybrid cross*—a cross that tracks the inheritance pattern of a single character. Apply the Law of Segregation.

   Example: In pea plants, tall (T) plants are dominant to short plants (t). There are three allelic combinations:
   - ➤ TT—homozygous dominant (true breeding)/Tall
   - ➤ Tt—heterozygous or hybrid/Tall
   - ➤ Tt—homozygous recessive (true breeding)/Short

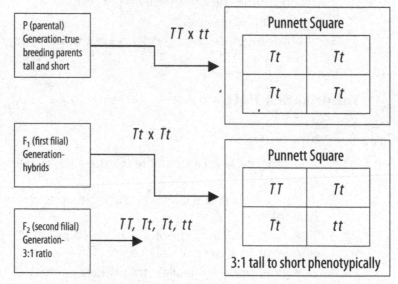

Cross two true breeding plants of tall and short

   ➤ Note: Punnett squares can be used to determine the genotypes and phenotypes of progeny from a genetic cross. Monohybrid or multi-hybrid crosses can be used.

2. *Test cross*—A cross that determines whether the dominant parent is homozygous dominant or heterozygous. Always cross the dominant parent to a homozygous recessive. Assume black (B) is dominant to white (b) for cat coat color.

Black parent could be BB or Bb. White parent is bb.

➤ If BB x bb, all progeny will be black carriers.

➤ If Bb x bb, ½ of the progeny are black and ½ are white.

3. *Dihybrid cross*—a cross between two different characteristics; demonstrates Law of Independent Assortment.

Example: In pea plants, tall (T) plants are dominant to short plants (t). Green leaf (G) is dominant to yellow leaf (g).

Cross two true breeding plants of tall green and short yellow

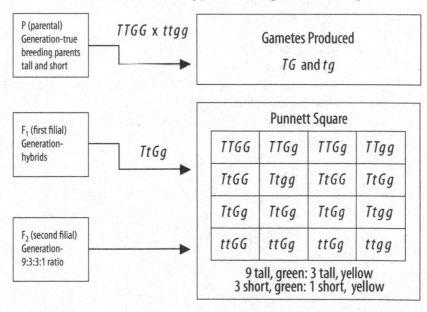

P (parental) Generation-true breeding parents tall and short	*TTGG* x *ttgg*	**Gametes Produced** *TG* and *tg*

Punnett Square

F₁ (first filial) Generation- hybrids — *TtGg*

F₂ (second filial) Generation- 9:3:3:1 ratio

TTGG	*TTGg*	*TTGg*	*TTgg*
TtGG	*Ttgg*	*TtGG*	*TtGg*
TtGg	*TtGg*	*TtGg*	*Ttgg*
ttGG	*ttGg*	*ttGg*	*ttgg*

9 tall, green: 3 tall, yellow
3 short, green: 1 short, yellow

E. Using the Laws of Probability in Genetics

1. *Probability (p)*—the number of times an event is expected to occur divided by the number of opportunities for the event to occur. A probability can be expressed as a fraction, a percentage, or a decimal; for example:
   i.   ¼ = 0.25 = 25%
   ii. ½ = 0.5 = 50%
   iii. 1/1 = 1 = 100%

2. *Law of Multiplication*—used to calculate the probability of independent events occurring; therefore, for genes that are linked the law of multiplication cannot be followed.

   Example 1: Assume the following cross: AaBbCc x AabbCC. What are the chances of the following progeny?
   (a) AabbCC
   (b) aabbCc
   (c) AAbbCC

   Answer: Perform each individual monohybrid cross and use the law of multiplication.
   Aa x Aa = 1/2 Aa, 1/4 aa, 1/4 AA
   Bb x bb = 1/2 Bb, 1/2 bb
   Cc x CC = 1/2 Cc, 1/2 CC
   (a) AabbCC = ½ x ½ x ½ = 1/8
   (b) aabbCc = ¼ x ½ x ½ = 1/16
   (c) AAbbCC = ¼ x ½ x ½ = 1/16

   Example 2: Assume the following genotype: AaBBCcddEeFf. How many different gametes are possible?

   Answer: Determine how many different gametes are possible for each set of alleles.

   $$\left.\begin{array}{l} \text{Aa} = 2 \text{ (either A or a)} \\ \text{BB} = 1 \text{ (only B)} \\ \text{Cc} = 2 \text{ (either C or c)} \\ \text{dd} = 1 \text{ (only d)} \\ \text{Ee} = 2 \text{ (either E or e)} \\ \text{Ff } = 2 \text{ (either F or f)} \end{array}\right\} \begin{array}{l} 2 \times 1 \times 2 \times 1 \times 2 \times 2 = \\ 16 \text{ different gametes} \end{array}$$

F. *Non-Mendelian Genetics*—genetics that do not follow the inheritance patterns of Mendel's initial pea plant experiments.

1. *Incomplete dominance*—the phenotype of the offspring has an appearance that is between that of both parents. This is not a blending hypothesis. The dominant allele is not fully expressed.

Snapdragons

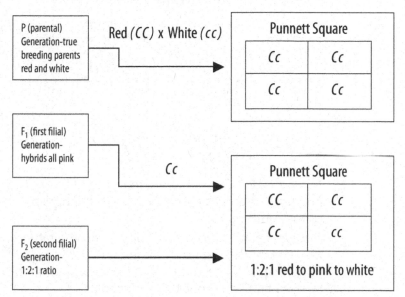

2. *Codominance*—both alleles are expressed at the same time.
   i. MN Blood system (M and N are blood group antigens found on the cell surface of a red blood cell).
   ii. There are three allelic combinations:
      ➤ *MM*—homozygous dominant (only produce M antigen on cell surface).
      ➤ *MN*—heterozygous (produce M and N antigens on cell surface).
      ➤ *NN*—homozygous recessive (only produce N antigen on cell surface).

3. *Multiple alleles*—many different alleles can control the expression of a character.
   i. *ABO Blood System*—carbohydrate antigens found on the cell surface.

Geno-type	Phenotype	Antigen on Cell Surface of Red Blood Cell	Antibodies Present in Blood
ii	O blood type	None	Anti-A, Anti-B
$I^A I^A$ or $I^A i$	A blood type	B	Anti-A
$I^B I^B$ or $I^B i$	B blood type	A	Anti-B
$I^A I^B$	AB blood type	AB	None

Example ABO Cross: Assume that a child has type B blood and the father was type A. What are the possible genotypes of the mother?

Answer: Child could be $I^B I^B$ or $I^B i$ and the father could be $I^A i$. The mother could be either $I^B I^B$ or $I^B i$ or $I^A I^B$. If the father was $I^A I^A$ no matter what genotype the mother is, a type B child could not be produced.

4. *Pleiotropy*—one gene causes multiple different phenotypic effects on an organism.
   ➤ An example of pleiotropy is PKU, which causes the following:

Human Disease PKU (phenylketonuria) Phenotypes		
Cognitive Impairment	Seizures	Skin Disorders

5. *Epistasis*—one gene affecting the expression of another gene. F_2 offspring phenotypic ratio is usually 9:3:4.
6. *Polygenic inheritance*—two or more genes affecting one phenotype. Examples include skin color and cancer, which

is the most common polygenic inherited disorder. Polygenic inheritance leads to a bell curve distribution of phenotypes.

G. *Pedigree Analysis*—a visual depiction of inheritance patterns in multiple family generations.

1. Basic Rules
   i. If two affected people have an unaffected child, it must be a dominant pedigree: [A] is the dominant mutant allele and [a] is the recessive allele. Both parents are Aa (hybrid carriers) and the unaffected child is aa.
   ii. If two unaffected people have an affected child, it is a recessive pedigree: [A] is the dominant allele and [a] is the recessive allele. Both parents are Aa (hybrid carriers) and the affected child is aa.
   iii. If every affected person has an affected parent, it is a dominant pedigree (no skipping of generations).
   iv. Dominant traits never skip generations, while recessive traits can skip.
   v. Squares are male.
   vi. Circles are females.
   vii. Mating is indicated by the connection with a line.

   viii. Filled-in circles or squares indicate an affected person.

   ix. *Sex-linked dominant*—all females descending from the affected males have the disease.
   x. *Sex-linked recessive*—no male carriers possible and skips generations.
   xi. *Autosomal recessive*—carriers are present, so skips generations. 50% males and females affected.
   xii. *Autosomal dominant*—no carriers or skipping of generations. 50% males and females affected.

H. Autosomal Genetic Disorders

Recessive Inherited Disorder—absence or malfunction of protein Must receive both nonfunctional copies from parents; therefore, affected individual is homozygous recessive (aa).	
Disease	**Outcome**
Albinism	Lack of pigment in the skin, eyes, and hair. May lead to skin cancers.
Cystic Fibrosis	Defective or absent chloride channel protein in membranes, causing a build-up of mucus in lungs. Person is prone to bacterial infections.
Tay-Sachs	Defective or absent lipase enzyme in brain. Predominant in Jewish population.
Sickle cell disease	Defective hemoglobin protein. Mostly affects the African-American population.

Dominant Inherited Disorder—absence or malfunction of protein Must receive at least one nonfunctional copy from one parent; therefore, affected individual is heterozygous (Aa) or homozygous dominant (AA).	
Disease	**Outcome**
Achondroplasia	Dwarfism
Huntington's Disease	Degenerative breakdown of the nervous system.

1. As researchers gain more knowledge about many of these genetic disorders, there are also numerous social, medical, and ethical issues surrounding these diseases. Tay-Sachs disease, for example, can lead to pre-conception screenings to determine the probability of the couple having a child with this disorder; it can also lead to social, medical, and ethical challenges as how to approach a pregnancy of a child diagnosed with Tay-Sachs.

2. There is also a civic issue surrounding genetic disorders. As companies begin to unveil the genetic makeup of many of these disorders and develop tests and treatments for them, who or what owns this knowledge? Can a gene or a test for a gene be patented? Do insurance companies have "rights" to prescreenings done by individuals who are at higher risks for carrying certain genes, like those that are linked to breast cancer?

I. Linked Genes—the alleles of two genes located on the same chromosome often do not show independent assortment; instead, they exhibit a special type of inheritance called *linkage*.

1. If two genes are located close together on the same chromosome, they do not assort independently because they are physically linked to each other.
2. *Example of Linked Genes Experiment*—Thomas Hunt Morgan performed genetic crosses with the fruit fly (*Drosophila melanogaster*) and used the following terms:
   i. *Wild type*—most common phenotype in the population.
   ii. *Mutants*—alternative phenotypes to the wild type.
3. Morgan performed the following dihybrid mating:
   Example: In fruit flies, gray (g+) body color is dominant to black body color (g). Normal wings (w+) are dominant to dumpy wings (w).
   Cross a double heterozygote to a double recessive (g+g w+w x ggww).
   Expected phenotypes of 1,000 offspring would be:
       250 wild type (gray normal)/parental phenotype
       250 black dumpy/parental phenotype
       250 gray dumpy/recombinant phenotype
       250 black normal/recombinant phenotype
   Observed phenotypes of 1,000 offspring were:
       450 wild type (gray normal) /parental phenotype
       450 black dumpy/parental phenotype
       50 gray dumpy/recombinant phenotype
       50 black normal/recombinant phenotype
   The high number of observed parental phenotypes indicated that the genes for body color and wings were linked to each other. Linked genes are on the same chromosome and are very close to each other. Linked

genes are inherited together and recombination between the genes is very low.

Calculation of Recombination Frequency or the measure of genetic linkage between 2 genes (also called map units).

$$\text{Recombination Frequency} = \frac{\text{\# of recombinants}}{\text{total offspring}} \times 100$$

Using the data above:

$$\text{Recombination Frequency} = \frac{100}{1000} \times 100 = 10\%$$

Only 10% of the time will there be recombination between the genes for body type and wings.

4. *Genetic Maps*
   i. Recombination Frequency allows you to create genetic maps that estimate the distance between genes.

   Example: Assume the following Recombination Frequencies. Determine the genetic map for genes W,X,Y,Z.

   W-Y, 7 map units
   W-X, 26 map units
   W-Z, 24 map units
   Y-X, 19 map units
   Y-Z, 31 map units

   Answer:

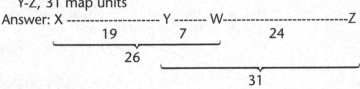

5. *Sex-linked Genes*—genes that are carried on the X-chromosome.
   i. Females carry two X chromosomes, XX.
   ii. Males carry 1X and 1Y, XY.
   iii. Inheritance patterns of sex-linked genes:
      ➤ A father will always transmit the sex-linked trait to his daughter. His son receives the Y and does not inherit the trait.
      ➤ Only females can be carriers of sex-linked traits. Therefore, a carrier female who mates with a normal male transmits the mutant allele to half of her sons

and half of her daughters. Examples of sex-linked traits include *hemophilia* and *muscular dystrophy*.

➤ *Barr body*—one of the female's X chromosomes is randomly inactivated in order to have the same gene dosage as males for sex chromosomes. The chromosome tends to look smaller in physical structure. Example of the phenotypic output of X-inactivation are *calico-colored cats.*

Practicing a variety of genetics problems is an essential test preparation activity for the AP® Biology exam. Some genetics problems provide information about parents and ask about phenotypic ratios of their progeny; others give information about progeny and ask about the parents' genotypes or phenotypes. Be prepared for both. Be sure to practice pedigree problems involving dominant, recessive, and sex-linked traits.

UNIT 6
GENE EXPRESSION AND REGULATION

DNA Structure and Replication

I. Key Concepts

A. *Deoxyribonucleic Acid (DNA)*—genetic material

B. The structure of a DNA molecule is the key to understanding how each strand of DNA can act as a template for the replication of the other strand during DNA replication and for the production of RNA during transcription.

C. DNA replication is a *semi-conservative process* that produces two new DNA molecules, each of which consists of one old strand and one newly synthesized complementary strand, and which are checked for errors by proofreading and repair processes.

II. Discovery of DNA as the Genetic Material

A. Transformation Experiments of Griffith, Avery, McCarty, and MacLeod.

1. Smooth (contains capsule) living *Streptococcus pneumonia* injected into a live mouse; it resulted in a dead mouse.
2. Rough (no capsule) living *Streptococcus pneumonia* injected into a live mouse; it resulted in a healthy mouse.
3. Heat-killed smooth (capsule destroyed) *Streptococcus pneumonia* injected into a live mouse; it resulted in a healthy mouse.
4. Heat-killed smooth (contains capsule) mixed with living rough (no capsule) *Streptococcus pneumonia* injected into a live mouse; it resulted in a dead mouse.

5. *Interpretation of the experiment*—DNA from the heat-killed smooth cells "transformed" the rough cells into smooth cells that killed the mouse. The transforming agent was DNA.

B. Hershey-Chase Experiment

1. Worked with T2 bacteriophage or a virus that infects bacteria.
2. Bacteriophage were radioactively labeled with P32 (DNA) or S35 (protein coat of bacteriophage).
3. When separate experiments were completed, it was found that bacteria contain the radioactively labeled P32 DNA of the bacteriophage.
4. *Interpretation of the experiment*—bacteriophage injected their DNA into the host bacterium to produce progeny phage, indicating DNA as the genetic material.

C. Watson and Crick

1. *James Watson, Francis Crick, Rosalind Franklin,* and *Maurice Wilkins* contributed to constructing the double helical model of DNA.
2. *X-ray crystallography*—a technique used to measure the shapes of molecules—was contributed by Franklin and Wilkins to determine that DNA was a double helix.

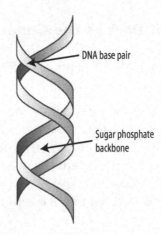

DNA base pair

Sugar phosphate backbone

3. Watson and Crick used this and other data to construct a *three-dimensional model.*

   i. Two strands of complementary DNA twist to form a *helix* often described as a spiral ladder with each base pair representing a single step.

   ii. Each turn of the double helix contains 10 base pairs and is 34 angstroms long.

   iii. The width of the helix is uniform and is 10 angstroms across.

4. Base pairing rules of purines and pyrimidines were established (also known as Chargaff's rule).

Hydrogen bonds indicated by dashed lines.

Adenine   Thymine

   i. Adenine (purine) pairs with thymine (pyrimidine). 2 hydrogen bonds for base pairing.

Guanine   Cytosine

   ii. Guanine (purine) pairs with cytosine (pyrimidine). 3 hydrogen bonds for base pairing.

D. Meselson-Stahl

1. Experiment indicated that replication of DNA is semi-conservative, or one old strand is used for the synthesis (template) of a new strand.

2. Experiment showed that both heavy and light nitrogen would be incorporated into the daughter DNA during the first round of DNA replication. In the second round of replication, daughter strands would have only light nitrogen since the heavy nitrogen was removed. Banding patterns indicated a semi-conservative model is favored over conservative or dispersive.

 III. **DNA Structure**

    A. The structure of a deoxyribonucleic acid (DNA) molecule is based on the pairing of nucleotides along the lengths of two complementary DNA strands, each of which has a sugar-phosphate backbone and twists to form a double helix.

    B. *Sugar-phosphate backbone*—of each DNA strand is a repeating chain of the 5-carbon sugar, *deoxyribose*, and a *phosphate group* composed of a phosphorus atom surrounded by 4 oxygen atoms.

        1. The sugar-phosphate chains are held together by *covalent bonds* that are generally only broken or formed by *enzymes*.

        2. The sugar-phosphate backbone of each strand is an identical feature of all DNA molecules.

    C. *Nitrogenous bases*—attached to each sugar is one of the four nitrogenous bases composed of carbon and nitrogen rings.

        1. *Purines*—adenine (A) and guanine (G), consist of two rings.

        2. *Pyrimidines*—thymine (T) and cytosine (C) have a single ring.

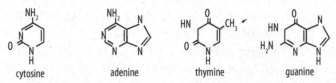

cytosine   adenine   thymine   guanine

    D. *Complementary pairing*—the complementary pairing of nitrogenous bases is the basis of the double-stranded structure of a DNA molecule.

1. *Base pairing* involves the formation of *hydrogen bonds* between the bases that are located toward the center of each DNA molecule, and they serve to hold the two strands together.
   i. Hydrogen bonds are weaker than covalent bonds, allowing the two strands of the DNA molecule to be separated for DNA replication and transcription.
   ii. A pyrimidine can only pair with a purine due to their sizes and the types of hydrogen bonds possible between the bases.
2. Base pairing occurs along the entire length of the two DNA strands and results in one strand being an exact complement of the other.
3. Knowing the sequence of one strand of DNA, the sequence of its complement can be deduced by the use of the base pairing rules.

   -----G A T T C G T A A G G C------ one strand of DNA
   -----C T A A G C A T T C C G------ complementary strand of DNA

An understanding of the similarities and differences between DNA replication, transcription, and translation is a popular topic for test questions.

IV. DNA Replication

A. During replication, each strand of a DNA molecule acts as a template for the synthesis of the other strand, and when errors do occur, proofreading and repair mechanisms keep mutation rates low.

B. All of the chromosomes of an organism's *genome* (all of an organism's DNA) are copied prior to each cell division.

1. Each strand acts as a *template* for the synthesis of its complementary strand through the addition of nucleotides to the growing end of the complement.

2. DNA replication begins at *origins of replication* distributed at many sites along the length of each eukaryotic chromosome and usually at a single site on the chromosome of prokaryotes.

3. *DNA polymerase* is the enzyme that catalyzes the addition of nucleotides to the ends of a growing strand of DNA through a process called elongation.

C. The process of DNA replication occurs as follows:

1. Enzymes, called *helicases,* unwind the helix at the origins of replication and help break the hydrogen bonds holding the strands together, creating a *replication fork*.

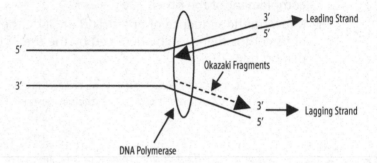

i. *Leading Strand*—the daughter strand that is synthesized into the replication fork. This strand is synthesized in a continuous fashion.

ii. *Lagging Strand*—the daughter strand that is synthesized away from the replication fork. This strand is synthesized in a discontinuous fashion or in fragments called *Okazaki fragments*.

2. An enzyme called *primase* then synthesizes a short segment of *RNA* called a *primer* that is complementary to nucleotides on the DNA strand.

3. *DNA polymerases* bind to each separated strand and begin adding the proper complementary nucleotides to the primer to produce a new copy of each strand.

4. The bond formed between two nucleotides is a covalent bond between the deoxyribose sugar of one and the phosphate of the other.

5. Another enzyme, called *DNA ligase,* helps seal gaps between the many growing strands by taking a 5' phosphate and 3'

hydroxyl and linking them together, thereby helping join the Okazaki fragments into a single strand.

6. DNA polymerases keep moving along the strands until synthesis of both strands is completed.

D. Each new molecule of DNA consists of one of the original DNA strands hydrogen bonded to its newly synthesized complement.

E. *Proofreading* during DNA replication and repair of damaged DNA results in low mutation rates at the nucleotide level.

1. DNA polymerase makes mistakes at a rate of about 1/10,000 base pairs, but proofreading and repair mechanisms reduce that rate to 1/1,000,000,000.

2. Errors are usually corrected by enzymes that move along the new DNA molecule and replace any base that has been mismatched.

3. DNA molecules are also susceptible to damage by chemicals or radiation and are repaired in a similar manner.

Major Enzymes and Proteins in DNA Replication		
Enzyme	**Substrate**	**Action**
DNA helicase	Double-stranded DNA	Opens up the DNA strand for replication
Single-stranded binding proteins	Single-stranded DNA	Binds single-stranded DNA and keeps replication fork open
DNA primase	Single-stranded DNA	Lays down an RNA primer on single-stranded DNA for DNA polymerase to hook up with
DNA polymerase	Single-stranded DNA	Adds the complementary base to the daughter strand using the parental template. Follows base pairing rules; adenine with thymine, guanine with cytosine
DNA ligase	Single-stranded DNA	Links a 5' phosphate with a 3' hydroxyl on the lagging strand

RNA Structure and Gene Expression

I. Key Concepts

A. RNA transcription produces rRNA, tRNA, and mRNA, all of which have different roles during the process of translation.

B. In its sequence of nucleotides, mRNA carries the genetic information present in an organism's DNA (also in the form of sequences of nucleotides) from the nucleus to the ribosome.

C. Translation (protein synthesis) converts the information in the nucleotide sequences of mRNA into information in the form of the amino acid sequences of proteins that are critical to the structure and functioning of cells.

II. RNA Compared to DNA

A. *Ribonucleic acid (RNA)* structure is similar to DNA in some ways, but has very important differences.

   1. RNA and DNA are both composed of nucleotides and both have strands consisting of a *sugar-phosphate backbone* with nitrogenous bases *attached to their sugars*.

   2. RNA and DNA differ in the following significant ways:
     i. The 5-carbon sugar in RNA's backbone is *ribose* instead of deoxyribose.
     ii. Instead of thymine, RNA uses the pyrimidine base, *uracil*.
     iii. RNA does not form a stable double helix along its entire length with a complementary strand of RNA, as DNA does.

> RNA is often present in a *single-stranded state*, but it also can base pair with DNA, with itself, and with other RNA molecules.

> When RNA does form base pairs through hydrogen bonding, guanine pairs with cytosine, and adenine pairs with uracil.

iv. The two DNA strands in double-stranded DNA are antiparallel in directionality.

3. The following figure shows the complementarity of DNA and RNA:

-----G A T T C G T A A G G C------ D N A
-----C U A A G C A U U C C G------ R N A

B. The functions of RNA and DNA are different.

1. *DNA stores genetic information* on how to make RNA and proteins, and it passes this information from cell to cell and from parents to offspring.

2. Different types of RNA function in different ways.

   i. *Messenger RNA (mRNA)* transfers genetic information from DNA in the nucleus to ribosomes in the cytoplasm, where its information is translated into proteins.

   ii. *Ribosomal RNA (rRNA)* is incorporated into large complexes called ribosomes, which are the sites of protein synthesis in the cytoplasm; rRNA also regulates gene expression at the level of mRNA transcription.

   iii. *Transfer RNA (tRNA)* carries amino acids to a ribosome so they can be assembled into proteins.

3. The three types of RNA have different structures and functions, but all are made during the process of transcription, and all are important in the subsequent production of proteins during translation.

III. Transcription

A. *Transcription*—the process whereby information contained in the nucleotide sequences of genes is transferred to RNA molecules.

B. Transcription has three basic steps and occurs in the nucleus of eukaryotes and the cytoplasm of prokaryotes:

1. *Initiation*—the initiation of transcription is controlled by interactions between various proteins and various regions of a gene.
   i.   During initiation, the enzyme RNA polymerase binds to the promoter and opens a portion of the gene to create a transcription bubble.
   ii. RNA polymerase begins to synthesize an RNA transcript complementary to only one strand of DNA called the template.
2. *Elongation*—during elongation, RNA polymerase moves along the template DNA, adding nucleotides to the elongating strand of RNA.
   i.   RNA polymerase catalyzes the formation of a covalent bond between each new nucleotide by joining the ribose of one nucleotide to the phosphate of the other.
   ii. Each incoming nucleotide base pairs with its complementary pyrimidine or purine in the DNA template sequence on the gene.
3. *Termination*—when a termination sequence, or terminator, at the end of the gene is reached, RNA polymerase leaves the promoter, and the RNA transcript is released.

C. In eukaryotes, before RNA leaves the nucleus, it is modified.

D. rRNA—as part of the ribosomal subunits—plus mRNA and tRNA all travel through the nuclear pores of the nuclear envelope to the cytoplasm where they all participate in protein synthesis.

IV. Translation

A. *Translation*—also called protein synthesis, occurs in ribosomes in the cytoplasm of prokaryotic and eukaryotic cells.

B. Synthesis of protein from mRNA occurs in the 5′ to 3′ direction.

C. Requires all 3 RNA molecules (mRNA, tRNA, and rRNA) and codons (sets of 3 nucleotide RNA bases that code for amino acids).

D. The process of translation has three major steps:

1. *Initiation*—the start codon AUG (calls for the amino acid methionine) on the mRNA transcript is recognized by the ribosome.
   i. A tRNA—which carries the amino acid methionine and has the anticodon UAC—attaches to the ribosome at one of the sites set aside for tRNA on the ribosome, and its anticodon hydrogen bonds to the start codon, AUG.
2. *Elongation*—once the binding occurs, the entire ribosome translocates down another 3 bases and reads another codon sequence, where another tRNA brings in the appropriate amino acid.
   i. A peptide bond between the amino acids is formed via an enzymatic reaction promoted by the rRNA portion of the ribosome.
3. *Termination*—occurs when one of the stop codons (UAA, UGA, UAG) is read and the protein is released from the ribosome.

E. Protein activities can, in turn, affect the phenotype of an organism. Comparison of normal proteins with proteins that an abnormal allele is coding for allows scientists to begin to determine possible courses of treatments, if any. For example, in the case of albinism, the colorless compound DOPA is not converted to melanins. For an organism without albinism, the reaction should look like this:

DOPA— — — — — —·→ melanins

In albinism, the reaction looks like this:

DOPA— — — —//— —→ (no melanins)

Bottom line: A protein contained an abnormal allele that could not produce a specific product that was necessary for this reaction to proceed, and it resulted in albinism.

Nucleic Acid Technology and Applications

I. Key Concepts

A. DNA technology is a collection of procedures for manipulating and analyzing DNA that aid in all aspects of biological research and in developing technical applications for a wide range of purposes.

B. DNA technology has created many powerful tools for basic research, as well as for commercial use in agriculture and medicine.

II. Genetic Engineering Techniques

A. DNA Cloning

1. *Restriction Enzymes*—used to cut DNA molecules at specific locations called restriction sites.
   Example: Restriction site

   ```
   -----ACTGGA----                    -----A    CTGGA----
   -----TGACCT----        ──────▶      -----TGACC    T----
   ```

2. *Recombinant DNA*—combining DNA sequences that would not normally occur together to form one piece of DNA. The enzyme DNA ligase is added to seal the strands together.

```
-----A      CTGGA---- +  ┌──────────┐          ─────▶  -----A ┌──────────┐ CTGGA----
-----TGACC      T----    └──────────┘  DNA Ligase       -----TGACC └──────────┘ T----

   Cut DNA           Foreign DNA                              Recombinant DNA
```

3. Cloning Vector—original plasmid that is used to carry foreign DNA into a cell and replicate there.

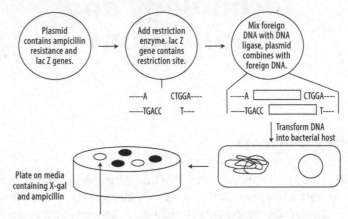

Bacterial clone that is white and grows on plate is clone containing recombinant DNA plasmid.

1. When the restriction enzyme is added to plasmid, *lac Z* is destroyed and nonfunctional. The *lac Z* gene produces the enzyme β-galactosidase, which breaks down the sugar X-gal causing the colony to appear blue. If the *lac Z* gene product is not made, the colony appears white; if the gene is functional, the colony appears blue.

2. Ampicillin resistance gene allows bacteria to grow in the presence of the antibiotic ampicillin.

3. Media is selective for clones that have the ampicillin resistance gene and differential for blue or white colonies.

4. Colony that is growing on plate (ampicillin resistance) and white are correct clones carrying recombinant DNA.

B. DNA Gel Electrophoresis

1. DNA is placed in a gel made of a polysaccharide called agarose or acrylamide (used for smaller fragments).

2. Migration of DNA is based on size differential of DNA fragments. An electric field is passed through DNA molecules and the molecules travel toward the positive end (cathode) due to negative charge of phosphate on DNA.

3. Larger molecules travel slower; smaller molecules travel faster.
4. Marker DNA of a standard size is used to approximate the size of unknown molecules. Marker is measured in kilobase pairs.
5. Visualization of DNA is done by staining the gel with ethidium bromide, which increases the visual difference between DNA and the gel.

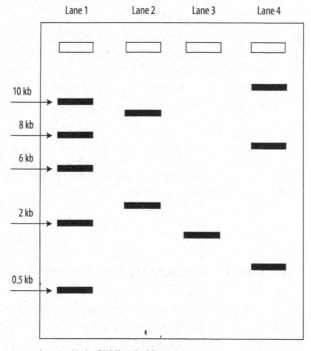

Lane 1 – Marker DNA/Standard Size
Lane 2 – 2 bands roughly 9 kb and 3 kb
Lane 3 – 1 band roughly 1.8 kb
Lane 4 – 3 bands roughly 12 kb, 7 kb, and 1 kb

Electrophoresis can be used for DNA and protein identification, isolation of different types of DNA or protein, calculating the size of fragments (DNA and protein), crime scene investigation, and genetic testing.

C. Polymerase Chain Reaction

1. A method to take a small amount of DNA and amplify (increase) the amount.
2. Based on progressive heating and cooling of DNA strands with the addition of primers and DNA polymerase.

D. DNA Fingerprinting

1. A technique used by forensic scientists to help determine the DNA of individuals.
2. The DNA of humans is highly homologous. There are sequences called *Short Tandem Repeats* (STRs). These repeats vary in length and size for each human, and therefore, can be used as identifying factors of humans.
3. STRs can be visualized using DNA gel electrophoresis.

Fully understanding the cloning process is considered a major concept in the AP® Biology course. You should understand how restriction enzymes and vectors are used in tandem to construct a recombinant plasmid.

III. Applications of Genetic Engineering

A. Genetic engineering techniques have created transgenic plants that are now used in agriculture to increase crop yields, reduce pesticide and fertilizer use, improve nutritional quality of grains, and create plants tolerant to extreme weather conditions such as drought.

B. Practical uses of DNA technology in medicine include production of vaccines and other pharmaceutical products.

1. Genetic analysis and transgenic organisms are used to create more effective vaccines that are less likely to cause disease than traditionally manufactured vaccines.

2. Cloning human genes in bacteria using expression vectors has resulted in large supplies of important medicines such as insulin to treat diabetes and interferons and interleukins to treat acquired immunodeficiency syndrome (AIDS).

C. Research applications of genetic engineering have been extensive. For example, cloning, RFLP analysis, and PCR and related chromosome mapping techniques have been used to map the entire human genome (as well as the genomes of many other organisms critical to basic research), creating computer databases that are widely available to researchers in all fields, including medicine, mathematics, engineering, computer technology, and other biology disciplines.

Regulation of Gene Expression

I. Key Concepts

A. *Gene Expression*—the transcription and translation of a gene into protein—is controlled by DNA sequences surrounding the coding region of a gene and by regulatory proteins that bind to these sequences.

B. Control over gene expression is important for determining when, and in which cells, a protein will be made.

C. Gene regulation in bacteria involves control over the transcription of operons, and in eukaryotes involves multiple levels before, during, and after transcription and translation.

D. *Regulatory proteins* provide both positive and negative control mechanisms for gene expression. They inhibit gene expression by binding to DNA and blocking transcription (*negative control*), and they can stimulate gene expression by binding to DNA and stimulating transcription (*positive control*) or binding to repressors to inactivate their repressor functions.

E. Some genes, like *ribosomal genes*, are always turned "on" and are continuously expressed.

II. Gene Regulation in Prokaryotes

A. In *prokaryotes*, such as *bacteria*, the single chromosome contains many genes that are organized into operons.

1. An *operon* contains a promoter, an operator, and a group of structural genes.
   i. Several *structural genes*, often coding for proteins involved in the same metabolic process, are under the control of a promoter and operator.
   ii. A *promoter* is the part of the operon to which RNA polymerase binds to begin transcribing the structural genes.
   iii. An *operator* is the part of the operon to which a repressor protein can bind to stop expression of the structural genes.
2. *Repression*—occurs when a regulatory protein, called a *repressor*, binds to the operator, thereby blocking RNA polymerase from transcribing the genes.
3. *Induction*—occurs when a substance, called an *inducer*, binds to the repressor protein, inactivates it, and keeps it from binding to the operator, thereby activating transcription of the genes.

Regulation of a Bacterial Operon: The Lac Operon

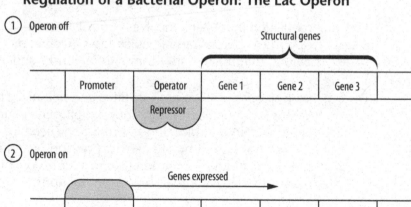

① Operon off

		Structural genes		
Promoter	Operator	Gene 1	Gene 2	Gene 3
	Repressor			

② Operon on

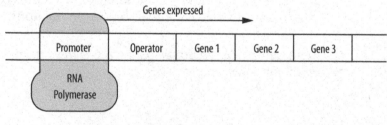

Genes expressed →

| Promoter | Operator | Gene 1 | Gene 2 | Gene 3 |

RNA Polymerase

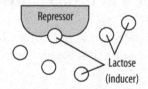

Repressor

Lactose (inducer)

4. The *lac operon* is induced by the sugar lactose.
   i. The structural genes of the *lac operon* control the utilization of lactose by the bacteria, *Escherichia coli* (*E. coli*).
   ii. When lactose is not present in its environment, *E. coli* has no need to turn on the *lac operon* because there is no lactose to metabolize; so the repressor is bound to the operator, and the genes are turned off.
   iii. If lactose becomes available, it acts as an inducer by binding to the repressor and inactivating it so that RNA polymerase can bind to the promoter and transcribe the structural genes.
   iv. When the structural genes are translated into proteins, the proteins help the bacteria use lactose as an energy source.
   v. As the lactose is broken down and used by the bacteria, eventually none is left to bind to the repressor to keep it inactive; so the repressor binds once again to the operator, thereby turning off expression of the structural genes.
5. In this way, bacteria save resources and energy by turning off an operon when its gene products are not needed.

Test Tip  *Know the basic parts of an operon and how operons are turned on and off.*

III. Gene Regulation in Eukaryotes

A. *Eukaryotic gene expression* is controlled at many levels as DNA's information is converted to mRNA and then into protein.

Opportunities for Control over Eukaryotic Gene Expression

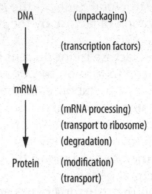

1. Only a small fraction of the genes in the genomes of multicellular organisms need to be expressed in any given cell type at any given time.
2. Eukaryotic genes are not organized into operons as in bacteria.
   i. Each gene has its own *promoter* and other *control elements* present in the DNA sequences surrounding the gene.
   ii. Many types of *regulatory proteins* bind to these elements to determine the timing and level of expression of any particular gene.
3. Controls over gene expression can occur before, during, or after transcription and translation.
   i. *Before transcription* can occur, the region of the genome in which the gene is located needs to be *unpackaged* to allow access to regulatory proteins.
      ➤ During interphase, some DNA remains as tightly packed as it was during mitosis and is called *heterochromatin*; genes in these regions of the genome are not able to be expressed.
      ➤ The uncoiling of DNA in a region—which is partly controlled by *histone proteins* associated with DNA—produces less tightly packed *euchromatin*, allowing regulatory proteins to access genes in that region.
   ii. To allow for *transcription* after a region of DNA has been unpackaged, regulatory proteins, such as *transcription*

factors, bind to the regulatory elements of the gene, thereby allowing access to the promoter by RNA polymerase.

➤ Some transcription factors bind directly to the promoter to assist the binding of RNA polymerase.

➤ Other transcription factors bind to DNA regions surrounding the gene, called *enhancers*, to further assist RNA polymerase binding.

➤ Once *RNA polymerase* binds to the promoter, transcription can occur.

➤ By controlling the synthesis and activity of transcription factors, a cell determines which gene will be expressed and when.

iii. *After transcription*, opportunities for control of gene expression can occur during *RNA processing*, which may involve the speed or types of splicing that occur to remove introns and join exons together to produce the mature mRNA.

iv. *Before translation*, the amount of mRNA sent to the ribosome can be controlled by how efficiently it is *transported from* the *nucleus* to the ribosome, or by how many *mRNA transcripts* are *degraded* along the way.

v. After translation, opportunities for further regulation of gene expression may involve *protein modifications* and *transport*, such as clipping off small parts of the protein, adding sugars to the protein, or reading signals located in the protein's amino acid sequence that determine where the protein is to be transported.

➤ Gene regulation in both prokaryotes and eukaryotes accounts for some of the phenotypic diversity of organisms, even though they may have similar genes.

Genetic Variation

I. Key Concepts

A. Many different types of mutations of nucleotides and chromosomes can occur during DNA synthesis and meiosis, resulting in different effects.

B. Mutations in somatic cells and germ cells can lead to the formation of cancer.

II. Mutations: Causes, Types, and Consequences

A. *Mutations during DNA Replication*—mutations of one or a few nucleotides in DNA are called point mutations and usually occur during DNA replication.

1. *Substitution mutations*—occur when a nucleotide has been altered or incorrectly paired during DNA synthesis, thereby changing it to another nucleotide.
   i. Substitution that occurs within a *coding region* of a gene may or may not cause a change in the amino acid in that position of the protein.
   ii. Some substitutions can result in *detrimental effects*, such as changes in blood protein hemoglobin that result in sickle cell anemia disease.
2. *Insertion or deletion point mutations*—occur when one or a few nucleotides are inserted or deleted.
   i. Insertion or deletion of groups of three nucleotides within a coding region of a gene may simply cause insertion or deletion of amino acids in a protein.

ii. Insertion or deletion of one nucleotide, or groups of nucleotides that are not divisible by three, result in *frameshift mutations*—causing the reading frame (the coding region) to shift; this mutation causes all the amino acids after the site of the mutation to also be altered.

iii. A frameshift mutation, especially near the start of the gene, almost always results in a completely defective protein.

B. *Mutations during Meiosis*—mutations occurring during meiosis (as opposed to those that occur during DNA replication) can involve parts of chromosomes or whole chromosomes.

1. *Mutations in chromosome structure*—where regions of DNA much larger than those involved in point mutations are involved—are due to chromosome breakage.

i. *Deletion*—occurs if a region of the chromosome (that does not contain a centromere) is broken and does not rejoin the chromosome.

ii. *Duplication*—occurs if a broken portion of a chromosome becomes incorporated into its homologous chromosome.

iii. *Inversion*—occurs if the broken portion of the chromosome may also be inverted and reattached to the same chromosome.

iv. *Translocation*—occurs if a portion of the chromosome is moved from one chromosome to a chromosome that is not its homologue.

2. *Mutations in chromosome number*—involve the loss or gain of whole chromosomes, or duplication of whole genomes (all of an organism's chromosomes).

i. *Nondisjunction*—the failure of chromosomes to separate properly during meiosis—can result in the production of cells with abnormal numbers of chromosomes.

➤ Nondisjunction can occur during meiosis I or meiosis II. When nondisjunction occurs during meiosis I, a complete tetrad of one pair of homologous chromosomes moves to one side of the cell. When nondisjunction occurs during meiosis II, a pair of sister chromatids fails to separate, pulling both chromatids to one side of the cell.

> ➤ Regardless of whether nondisjunction occurs during meiosis I or meiosis II, the result is that some gametes are missing one chromosome and others have two copies of that chromosome.
> ➤ A normal gamete and a gamete missing one chromosome (n − 1) join to create a zygote with a missing chromosome that is called a *monosomic.*
> ➤ A normal gamete and a gamete with an extra chromosome (n + 1) may join to create a zygote with an extra chromosome that is called a *trisomic.*

C. *Consequences of Mutations*—the effects of mutations run the gamut from neutral to lethal, but mutations also provide the raw material for evolution by natural selection.

1. Mutations can occur in somatic cells or germ cells.
   i. If a mutation occurs in a germ cell, it can be inherited because the products of germ cells are gametes.
   ii. If a mutation occurs in somatic cells, it is not inherited through sexual reproduction, but it may have other effects in the organism in which it occurs, such as the development of *cancer.*
2. Germ cell mutations that do not affect the fitness of an organism are known as *neutral mutations.*
   i. Neutral mutations result in neutral variation—variations between organisms that do not seem to affect evolutionary fitness.
   ii. Many, if not most, of the DNA differences between members of the same species, such as those revealed by DNA fingerprinting, may not affect an organism's fitness.
   iii. Neutral variation may be important during evolution because environmental conditions vary over time, and a variation that is neutral under one set of conditions may be beneficial under a different set of conditions.
   iv. Sometimes the variation can improve the evolutionary fitness of the individual under a certain set of conditions; for example, *sickle cell anemia* allows those with this condition to be less likely to contract malaria, and in parts of the world in which malaria is rampant, avoiding contracting malaria improves one's evolutionary fitness under those conditions.

3. *Polyploidy*—occurs in plants, can create *new species* in one or a few generations.

4. *Allelic variation*—germ cell mutations create the *allelic variation* that underlies *phenotypic variation*.
    i.   A mutation in the regulatory parts of genes, such as a promoter, can affect when, where, and how much of a protein is produced in different cells in a body.
    ii. A mutation in a gene is important as a regulatory component in processes such as metabolism, cell-to-cell communication, growth, development, etc., and can create significant differences between individuals and between species.
    iii. Mutations that alter the expression of proteins or their amino acid sequences can result in *lack of a protein* or in a *nonfunctional protein* that may cause *inherited diseases*.

5. *Lethal mutations*—some germ cell mutations can include point mutations or the loss of parts of chromosomes or whole chromosomes, and result in the death of an organism before birth.

6. *Cancer*—a combination of germ cell and somatic cell mutations are involved in the development of cancer.
    i.   *Tumor cells* undergo cell division much more often than normal cells because they lack control over the cell cycle and other cell growth processes.
         ➤ Cells normally undergo cell division during activities such as development, growth, maintenance, and repair of an organism's body.
         ➤ Some cells divide repeatedly to produce masses of cells called *tumors*.
         ➤ Tumors become dangerous if they interfere with normal body functions, and if they spread—*metastasize*—to multiple locations within the body, causing *cancer*.
    ii. Mutations that cause tumors can occur in germ cells or somatic cells.
         ➤ A mutation in a germ cell can be inherited, and people with inherited mutations are more susceptible to developing some types of cancer.

➤ Mutations in somatic cells add to the effects of inherited mutations, making the development of cancer more likely.

iii. There are several *types of genes* that, when mutated, can result in cancer.

➤ Genes that normally regulate the cell cycle or a cell's response to growth hormones—called *proto-oncogenes* in this context—mutate to form *oncogenes*.

➤ Other normal genes, called *tumor-suppressor genes*, exert negative control over cell division processes, and when they are mutated, they may fail to keep cell growth under control.

➤ The *accumulation* of mutated genes over an organism's lifetime can turn otherwise normal cells into cancerous cells.

iv. *Exposure to carcinogens and mutagens* is the most likely cause of the *somatic mutations* that contribute to cancer development, but mutagens may also cause germ cell mutations that are carried by gametes.

➤ *Carcinogens* are substances in the environment that increase the risk of cancer.

➤ Most carcinogens are mutagens.

➤ *Mutagens* are substances that cause mutations. Examples include tobacco, asbestos, x-rays, ultraviolet light, and a host of other chemical substances in the environment.

v. Some *viruses* can contribute to cancer development by transferring oncogenes to host cells, or causing mutations in proto-oncogenes or tumor-suppressor genes of host cells.

Test Tip

Know the names, descriptions, and effects of the different types of mutations.

Genetic Variation in Prokaryotes and Viruses

A. Prokaryotes reproduce *asexually*, but they also have several methods of genetic recombination.

1. *Transformation* occurs when a prokaryote takes up foreign DNA from its environment.

2. *Transduction* is when a virus transfers prokaryotic DNA from one cell to another.

3. *Conjugation* occurs when a plasmid is transferred from one prokaryote to another through a special tube-like structure called a *pilus*.

B. Because prokaryotes have a very short generation time, mutation and genetic recombination play important roles in producing and maintaining genetic diversity.

C. Virus replication involves invading a host cell and eventually living off of the host by taking over the metabolic machinery (parasitic).

1. Viruses cannot reproduce independently.

2. Viruses attach to the host via cell surface receptors and inject their DNA into the host.

3. During the *lytic cycle*, which is a type of viral reproduction, the virus eventually kills the host.

4. During the *lysogenic cycle*, the virus replicates its genome without killing the host and forms *prophage* (incorporation of the viral DNA into the host chromosome).

D. Viruses are efficient at rapid evolution and acquiring new phenotypes. The wide variety of possible hosts, and in turn, the enormous possibilities of chromosomes and their respective genes through which a virus can replicate itself all contribute to viruses' abilities to possess the potential for expansive genetic variations.

E. *HIV*, the causative agent of acquired immunodeficiency syndrome (AIDS), is a retrovirus; it uses an enzyme called reverse transcriptase to synthesize DNA from an RNA strand and infects T4 helper cells.

F. *Prion* is the protein infectious particle or misfolded protein that converts other normal proteins into mutant form; it is also the causative agent for "mad cow disease."

IV. Variation Due to Sexual Reproduction

A. Sexual reproduction produces genetic variation in three ways: independent assortment of homologous chromosomes, crossing over, or random fertilization.

B. The first two genetic recombination events occur during meiosis I.

1. *Independent assortment of homologous chromosomes*—occurs during metaphase I and anaphase I, creating a variety of outcomes (gametes) that contain different combinations of an organism's maternal and paternal chromosomes.

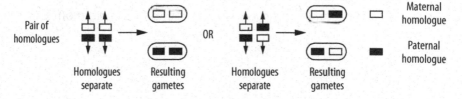

Depending on how homologues align during meiosis I, gametes with different combinations result.

2. *Crossing over*—occurs between homologous chromosomes during prophase I, creating entirely new chromosomes on which the organism's maternal DNA is mixed with his or her paternal DNA so that newly created chromosomes may be passed on to offspring.

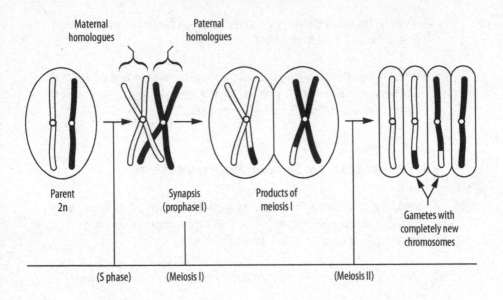

Maternal homologues   Paternal homologues

Parent 2n

Synapsis (prophase I)

Products of meiosis I

Gametes with completely new chromosomes

(S phase)   (Meiosis I)   (Meiosis II)

C. *Random fertilization* creates further genetic variation.

1. One sperm, out of the large variety of sperm a male can produce, joins with any of the large number of different eggs a female can produce.

2. This creates a large variety of different possible offspring.

UNIT 7

NATURAL SELECTION

Natural Selection and Evolution

I. History

A. Contributing Ideas to Darwin's Descent with Modification

1. In the 1700s and 1800s, the biological sciences were defined in terms of *natural theology* rather than scientific data and extrapolation. Several scientists began to use data to debunk natural theology as a means for explaining scientific findings.

2. Charles Darwin built on the ideas of other scientists to develop his theory of "descent with modification" by natural selection.

3. While examining the fossil record, both Jean-Baptiste Lamarck and Darwin agreed that species evolve over time, but each proposed a different mechanism:

   i. Lamarck proposed the (incorrect) mechanism for evolution—called the *inheritance of acquired characteristics*—which asserted that if a trait is used, it will be passed down to the next generation, but, if not used, then it will be discarded and not passed along. His theory is notable because of its emphasis on an organism's adaption to the environment.

   ii. Darwin also recognized that species change over time, but he proposed a *different mechanism* for how that change occurs, which he called natural selection.

4. Darwin was influenced by the ideas proposed by a number of other scientific thinkers, as well as his own extensive observations of biogeography and of plant and animal breeding.

      i. Charles Lyell, a geologist, proposed that the Earth had been around for a long period of time, that geological processes—such as volcanic eruptions—that occur currently also occurred in the past (*uniformitarianism*), and that these types of processes, over a long period of time, account for large-scale changes in the Earth's physical characteristics (*gradualism*).

    ii. These ideas led Darwin, and others, to conclude that the *strata* (and their fossils), observable in the exposed rock, represent distinct time periods during the Earth's history.

  iii. Thomas Malthus proposed that population size remains fairly steady, despite its capacity for exponential population growth because of disease, wars, and limited resources.

       ➤ Darwin felt that this situation applied more generally to all species and further proposed that the availability of limited resources led to competition between members of a species.

B. Natural Selection

1. By studying 12 different types of finches on the Galapagos Islands, Darwin made a link between the origin of a new species and the environment in which these species reside.

2. *Theory of Natural Selection*—reproductive success of an organism depends on its ability to adapt to the environment in which it resides. For example, several of the finches in the Galapagos Islands adapted their beak structure to find food.

3. Postulates of Natural Selection—

    i. Every population has variation in phenotypes due to genetic mutations.

   ii. If the environment cannot support the individuals who occupy it, then competition occurs between members of a species and affects the production of offspring. If an aspect of the environment causes some organisms to survive over others, it is said to exert *selective pressure*.

iii. Survival of individuals within a population will depend on their phenotype. Individuals with traits that promote survival will be more likely to pass these traits to offspring, allowing them to survive to reproduce and thus be more "fit." The traits that make them more "fit" are said to give them a selective advantage.

iv. Over time, the fittest organisms will survive, hence "survival of the fittest," and therefore more individuals in the population will be born with the advantageous phenotype.

> ➤ The result of natural selection is the adaption of populations to their environment, thus giving them a competitive advantage to survive.

v. A genetic variation, such as the average beak length of a finch that changes based on the year, is an example of adaption. A finch population has a variety of beak lengths (phenotypes) due to genetic mutations that exist in the population. Therefore, from the data that follows, we see that during some years, the average beak length of the finches is longer. This was when there was less rain (1950 and 1980) and seed sizes change due to the amount of rain. The long-beaked finches that survive in 1950 and 1980 are able to eat seeds that are available in the dry season. A larger beak indicates a selective advantage. However, when the rainy season comes, different seeds are available, and finches with smaller beaks are better able to get food in order to survive and reproduce.

➤ Below is hypothetical data: Dry seasons are 1950 and 1980; wet seasons are 1960, 1970, 1990, and 2000.

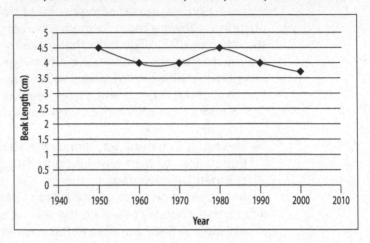

Darwinian biology permeates all aspects of biology. Early theories of evolution can make up a series of questions on the AP® Biology exam. Be sure to familiarize yourself with the major evolutionary theories for the exam. Remember that the term "fitness" means an organism's ability to survive to reproduce.

Evolution: An Ongoing Process

I. Evolution: An Ongoing Process

A. *Population Genetics*—study of genetic variation within a population of individuals.

1. *Population*—a group of individuals in the same geographical area that belong to the same species and are capable of interbreeding.
2. *Gene Pool*—the total sum of alleles within a population at a given time.

B. *Hardy-Weinberg Equilibrium*—a model for the study of the gene pool of a nonevolving population.

1. Hardy-Weinberg equilibrium indicates that the frequencies of two alleles do not change from generation to generation; a population is said to be in Hardy-Weinberg equilibrium if the following five conditions are met:
   i.   A very large population sample
   ii. No migration of individuals into or out of the population
   iii. No mutation
   iv. No natural selection
   v.   Random mating
   Note: These conditions are rarely met, but they provide a theoretical basis to test against.
2. To determine if a population is in Hardy-Weinberg equilibrium, allele frequencies must remain constant over generations. To determine allele frequencies, use the following equation.
$$p^2 + 2pq + q^2 = 1$$

p^2 = frequency of AA
q^2 = frequency of aa
$2pq$ = frequency of Aa
➤ Keep in mind: the combined gene frequency must be 100% so that $p + q = 1$.

Sample Problem #1: Assume a population of 500 pea plants in which green is dominant to yellow. Use the chart below to see how to calculate the frequencies of all phenotypes.

A = green, a = yellow

Phenotype	Green	Green	Yellow
Genotype	AA	Aa	aa
Number of pea plants (total = 500)	320	160	20
Genotypic frequencies	320/500 = 0.64 AA	160/500 = 0.32 Aa	20/500 = 0.04 aa
Number of alleles in gene pool	320 x 2 = 640 A	160 A + 160 a = 320 A & a	20 x 2 = 40 aa 40 a
Allelic frequencies	640 A + 160A = 800 A 800/1000 = 0.8A p = frequency of A = 0.8		160a + 40aa = 200 a 200/1000 = 0.2 a q = frequency of a = 0.2

➤ $p^2 + 2pq + q^2 = 1$
  • p^2 = frequency of AA = 0.8 × 0.8 = 0.64 = 64%
  • $2pq$ = frequency of Aa = 2 × 0.8 × 0.2 = 0.32 = 32%
  • q^2 = frequency of aa = 0.2 × 0.2 = 0.04 = 4%
➤ $p + q = 1$
  • 0.8 + 0.2 = 1 (Always check to make sure these numbers equal 1.)

Sample Problem #2: Assume that in a population of insects, body color is being studied: 36% of the insects represent the orange color, which is recessive, and 64% represent the black dominant phenotype.

> If each successive generation maintains the allele frequency, the population is said to be in Hardy-Weinberg equilibrium.

1) Determine the allelic frequencies.
2) Determine the genotypic frequencies.
   i. The recessive phenotype is key to this problem because the dominant represents both AA and Aa. However, recessive is *only* represented by aa. Use logic that q^2 = aa; therefore, the square root of .36 or q = 0.6. Since $p + q = 1$, $p + 0.6 = 1$, then $p = 0.4$.
   ii. Allelic frequencies are A = 0.4, a = 0.6
   iii. Genotypic frequencies follow the equation
   $$p^2 + 2pq + q^2 = 1$$
   iv. $p^2 = (0.4)^2 = 0.16 = 16\%$
       (AA or homozygous dominant) $\rightarrow$ Black phenotype
       $2pq = 2 \times 0.6 \times 0.4 = 0.48 = 48\%$
       (Aa, heterozygous dominant) $\rightarrow$ Black phenotype
       $q^2 = (0.6)^2 = 0.36 = 36\%$
       (aa, homozygous recessive) $\rightarrow$ Orange phenotype
       16 + 48 + 36 = 100 (Always double-check your numbers.)

Test Tip

Know the Hardy-Weinberg equilibrium concept and how to use it. It is highly likely that there will be questions on the AP® Biology exam that refer to it.

C. *Microevolution*—the change in the frequencies of alleles or genotypes in a population from generation to generation (evolution on a small scale) usually occurs if any of the five conditions of Hardy-Weinberg equilibrium are *not* met. Below are some ways the allele frequency can be changed.

1. *Genetic Drift*—defined as changes in the gene pool due to chance because of a small population. The small population directly contrasts the large population needed to maintain Hardy-Weinberg equilibrium.
   i. Causes a significant genetic change (microevolution) of a species if only a few members of a population migrate to find a new population.
   ii. Causes genetic change (microevolution) anytime a species is reduced to very small numbers due to chance events, such as hurricanes, earthquakes, fires, or habitat destruction.
2. *Bottleneck Effect*—a type of genetic drift that places stress on the gene pool due to some type of disaster, or massive hunting that inhibits a portion of the population from reproducing. The small population directly contrasts with the large population needed to maintain Hardy-Weinberg equilibrium.
3. *Founder Effect*—a type of genetic drift that occurs when a new colony is formed by a few members of a population. The smaller sample size narrows the genetic variation of the population. The small population directly contrasts with the large population needed to maintain Hardy-Weinberg equilibrium.
4. *Gene Flow*—transfer of alleles from one population to another through migration. The gametes of fertile offspring mix within a population, providing genetic variation. Genetic variation directly contrasts the no gene flow postulate needed to maintain Hardy-Weinberg equilibrium.
5. *Mutation*—a change in the genetic makeup of an organism at the DNA level. Mutation directly contrasts the no mutation postulate needed to maintain Hardy-Weinberg equilibrium.
6. *Non-random Mating*—individuals mating with those in close vicinity. Non-random mating directly contrasts with the random mating postulate needed to maintain Hardy-Weinberg equilibrium.
7. *Natural Selection*—reproductive success of organisms depends on their ability to adapt to the environment in which they reside. Natural selection directly contrasts with

the no natural selection postulate needed to maintain Hardy-Weinberg equilibrium.

➤ Keep in mind that any *genetic variation within a population can increase that population's genetic diversity*, even within the same species.

➤ Some *phenotypic variations can significantly increase or decrease the fitness of an organism* and the overall population.

   Examples include DDT resistance in insects, the peppered moth, and sickle cell anemia.

➤ *Humans can also impact other species* through: loss of genetic diversity within a crop species, overuse of antibiotics, and artificial selection.

D. *Speciation*—the origin of new species (a population of individuals who can mate with each other and produce viable offspring). In other words, when a population emerges that can no longer mate and produce fertile offspring with the original population, then a new species is said to have developed.

1. How Does It Occur?
   i. *Allopatric Speciation*—populations are separated by geographical isolation; thus, a new species can be formed following adaption to new surroundings.
   ii. *Adaptive Radiation*—Evolution of many species from a common ancestor. The finches Darwin found on the Galapagos Islands are an example of adaptive radiation.
   iii. *Sympatric Speciation*—populations are not separated by geographical isolation, but a new species is formed within the parent populations.
      ➤ *Autopolyploidy*—meiotic error causes a species to have more than two sets of chromosomes. Contribution is from one species.
      ➤ *Allopolyploidy*—polyploidy is a result of two different species.

2. How Fast Does It Occur?
   i. *Gradualism*—species are produced by slow evolution of intermediate species.
   ii. *Punctuated Equilibrium*—speciation occurs quickly at first and then is followed by small changes over a long period.

Test Tip

On the AP® exam, the definition of "species" is used in many questions. If asked to ascertain whether two organisms are from the same species, you must be able to determine whether they can mate and produce fertile offspring.

E. *Modes of Natural Selection*—natural selection can favor the allelic frequency in three ways. Below is an example of a bell curve or normal distribution of variations in a population. Three types of selections can take place that shift the bell curve to different frequencies (hence, evolution is taking place).

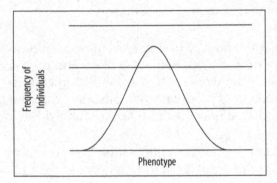

Original frequency of individuals shows a normal "bell-curve" distribution

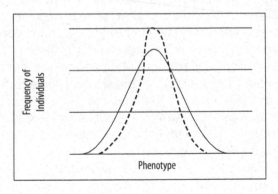

Stabilizing Selection—Extreme phenotypes are removed and more common phenotypes are selected

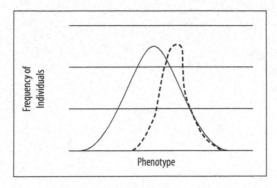

Directional Selection—One of the extreme phenotypes is selected

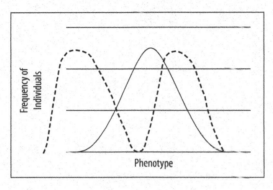

Diversifying Selection—Both of the extreme phenotypes are selected

There may be easy mathematical calculations on the AP® Biology test. For example, you should know how to apply the Hardy-Weinberg equation: $p + q = 1$, $p^2 + 2pq + q^2 = 1$. Also make sure you understand how to use the chi-square equation—don't memorize it. If needed, it will be provided:

$$X^2 = \Sigma\left[\frac{(o - \theta)^2}{\theta}\right].$$ *This statistical test can be used to determine whether significant change in allele frequency has occurred.*

Common Descent

I. Key Concepts

A. Evidence for Evolution

1. *Biogeography*—study of organisms and how they relate to the environment. Some organisms may be unique to certain geographies; hence, those organisms have adapted to live in that environment.

2. *Fossils*—help indicate the progression of organisms from simple to complex. For example, transitional fossils are fossils of animals that display a trait that helped the organism attain a competitive advantage. At one time, for example, whales had limb-like appendages indicating they may have been land dwellers.

3. *Comparative Anatomy*—study of anatomical similarities between organisms.
   i. *Homologous structures*—structures in organisms that indicate a common ancestor. For example, a human arm, cat leg, whale flipper, and bat wing all have a similar structure but different functions.
   ii. *Vestigial organs*—remnants of structures that were at one time important for ancestral organisms.

4. *Comparative Embryology*—comparing the embryonic development of one organism to another.

5. *Molecular Biology*—used in the study of evolution by looking at homology in DNA and protein sequences and genes; this study allows for an even broader level of comparison between organisms as different as prokaryotes, plants, and humans.

➤ Organisms share conserved core processes, which signal their evolution from a common ancestor and how widely distributed these processes have become among different species.

➤ *Examples:* DNA and RNA are carriers of genetic information through transcription, translation, and replication; the genetic code of many organisms is shared and is evident in many modern living systems; and many metabolic pathways, like glycolysis, are conserved.

➤ Structural evidence, such as cytoskeletons, membrane-bound organelles, linear chromosomes, and endomembrane systems, suggest that all eukaryotes are related.

B. Evolution Continues to Occur

1. Scientific evidence supports the premise that evolution continues to occur.
   Examples include:
   ➤ Emergent diseases
   ➤ Chemical resistances caused by mutations, such as resistance to antibiotics, pesticides, herbicides, and chemotherapy drugs
   ➤ Phenotypic change in a population (such as Darwin's finches in the Galapagos)
   ➤ Eukaryotes eventual development of structures such as limbs, brain, and immune system

C. Phylogenetic Trees and Cladograms

1. Represent traits that are either derived or lost due to evolution, such as opposable thumbs, the absence of legs in some sea animals, and the number of heart chambers in animals.
2. Illustrate that speciation has occurred and show when two groups were derived from a common ancestor.
3. Can be constructed from either morphological similarities or from DNA and protein sequence similarities by utilizing a computer program that measures the organisms' interrelatedness.

4. Provide a dynamic snapshot that is constantly being revised.
   Example

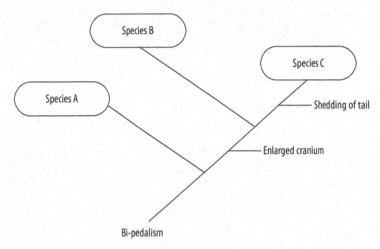

- ➤ Species B and C are more closely related to each other than to species A.
- ➤ All species are generated from an ancestor species with bi-pedalism.
- ➤ All species retain traits from the ancestor but have evolved to gain some specific trait through time.

Organism	Bi-pedal	Large Cranium	Tail Loss
Species A	X		
Species B	X	X	
Species C	X	X	X

Origin of Life

I. Key Concepts

A. The Origin of Life: Hypotheses and Evidence

1. Primitive Earth was thought to have the following atmospheric molecules—water (H_2O), methane (CH_4), hydrogen (H_2), and ammonia (NH_3)—and no oxygen.

2. These inorganic precursors of organic molecules on primitive Earth could have been formed as a result of an electrical spark and the lack of oxygen.

3. As a result, crude organic molecules including mono-saccharides, fatty acids, amino acids, and nucleic acids were formed.
   Miller-Urey experiment—tested the Oparin-Haldane model; the atmosphere on primitive Earth was the precursor for the synthesis of organic molecules.

4. Molecules then became the building blocks or monomers of more complex molecules.

5. Monomers began joining to form polymers that, over time, began to replicate, store, and transfer information. For example, amino acids are natural monomers that combine to form proteins, and nucleotides bond together to form nucleic acids.

6. Complex reactions could have occurred in a solution, known as the *Organic Soup Model*, or as reactions on solid reactive surfaces.

7. *RNA (ribonucleic acid)*—the first genetic material; it is capable of self-replication and can act as both genotype and phenotype. Eventually, *DNA (deoxyribonucleic acid)* became

the genetic material because of its stability over RNA and its ability to correct mutations.

B. Earth's History

1. Geographical Evidence:
   i. Earth is most likely around 5 billion years old.
   ii. Earth's environment was too hostile for life until about 3.9 billion years ago.
   iii. Earliest fossil records date back 3.5 billion years ago.
2. Molecular and Genetic Evidence:
   i. *Anaerobic prokaryotes* emerged approximately 4 billion years ago and represent the first origins of life.
   ii. Earliest living organisms were unicellular, had a genetic code, and were able to evolve and reproduce. The first organism is often known as LUCA, the last universal common ancestor.
   iii. Ribosomes, a cell membrane, cytoplasm, and DNA are common to all forms of life. Prokaryotes diverged into two types—bacteria and archaea—about 2.5 billion years ago.
   iv. Oxygen accumulated in the atmosphere approximately 2.5 billion years ago as a result of photosynthetic bacteria.
   v. Eukaryotes emerged 2 billion years ago via the *Endosymbiotic Theory*.
   vi. Until 500 million years ago, life was confined to aquatic environments. Plants eventually found a foothold on earth (root system) via a symbiotic relationship with fungi.
   vii. All living things come from a common ancestor.

C. Extinction and Adaptive Radiation

1. Extinction of a species is very common, and more than a dozen mass extinctions have occurred throughout geological history.
   i. Extinction rates become rapid during ecological stress.
   ii. For example, during the *Cretaceous extinction*, which occurred approximately 65 million years ago, about 50%

of species, including almost all of the dinosaurs, became extinct.

Test Tip

The names and dates of the major extinctions will not be on the AP® Biology exam; however, be prepared to use data to determine that extinction has occurred.

2. Adaptive radiation—the rapid development of new species from a common ancestor; may occur after a significant genetic change in a member of a species, or after a new habitat becomes available due to extinction of another (or many) species.
   i. Adaptive radiation causes an increase in speciation.
   ii. Occurs after mass extinctions.
   iii. Some significant adaptive radiations include the radiation of flowering plants after the development of effective dormancy and dispersal strategies (e.g., pollen and seeds) and the adaptive radiation of mammals after the mass extinction of dinosaurs.

UNIT 8

ECOLOGY

Population Dynamics

I. Key Concepts

A. Populations may experience *exponential growth* if there are no limiting factors in their environment and *logistic growth* if there are limiting factors.

B. The smaller a population is, the more likely it is to become extinct.

C. Worldwide, the human population is currently experiencing exponential growth, but is expected to begin to level off around the end of the 21st century.

II. Individuals and Populations

A. Individual organisms have mechanisms to withstand physical changes in their immediate environments.

B. The environment of an organism includes biotic and abiotic factors.

1. *Biotic* factors include all living organisms within the environment.
2. *Abiotic* factors are the physical factors of the environment and include temperature, precipitation, humidity, wind, salinity, and availability of oxygen, nutrients, and sunlight.

C. As abiotic factors change over time, or from place to place, in an organism's environment, the organism may respond in a variety of ways.

1. A *tolerance curve* describes how able or active an organism is over the range of change it may experience for a particular factor in its environment; at the extreme limits of its range, an organism may not survive.

Temperature Tolerance Curve

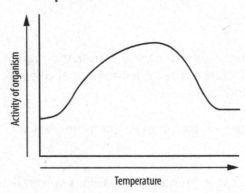

2. An organism may *acclimate* (adjust its tolerance) to an environmental factor, such as when humans produce more red blood cells as their bodies adjust to higher elevation.

3. *Regulators* are organisms that spend metabolic energy to internally regulate a physical factor, such as temperature or salinity, to keep it within a limited range even though their environment may exhibit a wider range for that factor.

4. In *conformers*, the factor is not internally regulated; instead, the conformer's internal factor changes to match the environmental factor as it increases or decreases.

Regulators and Conformers
Respond Differently to Environmental Change

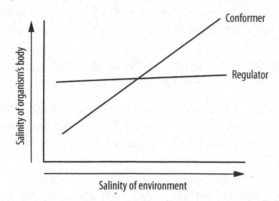

5. Individuals may respond to environmental change by temporary or permanent escape.
   i. An individual may move from location to location during the day, or *migrate* to another location.
   ii. An individual may become *dormant* through one of the following mechanisms.
      ➤ Some organisms may experience a period of inactivity called *torpor: hibernation* in cold weather or *estivation* in hot weather.
      ➤ Some organisms have resistant forms such as *spores* or *seeds*.

D. A *niche* refers to all of the roles of a species within its environment, including the biotic and abiotic features of its environment.

   1. The *fundamental niche* of a species is the total range of environmental factors it can tolerate and the total range of resources it can potentially use.
   2. The *realized niche* of a species is the actual extent to which it tolerates and uses its potential environment, factoring in how resources are often reduced by biotic factors such as competition with other organisms.
   3. *Generalists* are species with very broad niches, whereas the niches of *specialists* are more specific and limited.

III. Density and Dispersion

A. Population—group of individuals of the same species living together in the same location during the same period of time.

B. Population characteristics include size, density, patterns of dispersion, and age structure.

1. *Population size*—the number of individuals in the population; can be measured by direct counting in small populations or by sampling a portion of the population if it is larger.

2. *Population density*—refers to the number of individuals in a defined unit of space, such as the number of single-cell algae per milliliter of pond water or ferns per square kilometer of forest floor.

3. *Dispersion*—the pattern of distribution of individuals within a population.

   i. *Uniform* (or *even*) dispersion pattern—one in which the members of the population are spaced at relatively equal distances from one another, and often occurs in species that defend a defined territory.

   ii. *Random* dispersion pattern—each individual's position is independent of the locations of other individuals; for example, the dispersal of a plant's seeds by wind may result in the random location of the plant's offspring.

   iii. *Clumped* distribution pattern—the most common with most organisms in the population preferring to aggregate in the same area(s).

      ➤ Clumping can result from uneven distribution of the resources needed by the population's members.

      ➤ Clumping can also be the result of social behaviors that lead to swarming, flocking, or schooling among animals.

Dispersion Patterns

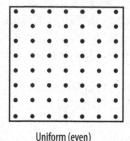 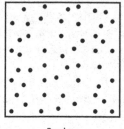 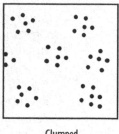

| Uniform (even) | Random | Clumped |

4. The *age structure* of a population is dynamic and changes over time due to varying birth rates, death rates, and life expectancies.
   i. A population with a greater number of younger, reproductively active members is expected to increase in size more rapidly than a population with fewer young individuals.
   ii. If life expectancy increases, the number of older members in a population is expected to increase.

Age Structure in Two Different Populations

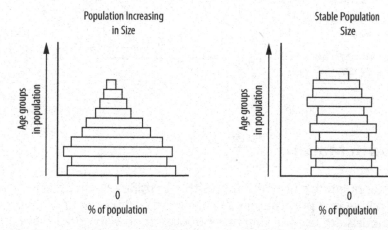

5. A *survivorship curve* shows the expected mortality (death) rates of members of a population over their potential life span.

    i. A *Type 1* curve is characteristic of species that have
few young and invest a lot of energy caring for them.
Survivorship is high for early and midlife individuals, but
drops precipitously with advanced age, indicating that
most members of the population live out their potential
maximum life span.

    ii. *Type II* curves describe populations in which the
members have more or less the same chance of dying
regardless of age.

    iii. A *Type III* curve is characteristic of species that produce
large numbers of offspring, most of which die before
reaching maturity.

Age Structure in Two Different Populations

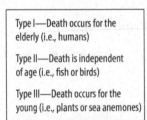

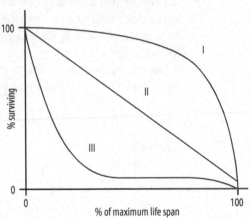

IV. Growth Models

A. A population's growth rate (birth rate minus death rate) is the
change in a population's size per a defined unit of time; two
models are used to describe population growth under different
conditions.

    1. *Exponential Model*—predicts the unlimited growth of a
population because of no limitation on resources; the result
is a J-shaped curve.

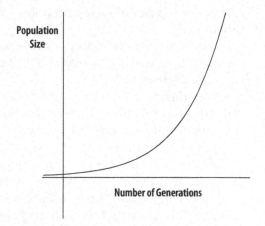

2. *Logistical Model*—more applicable to most populations in that it takes into account limited growth of a population due to limited resources; the result is an S-curve. The carrying capacity is the maximum population size a habitat can hold (defined by the letter K).

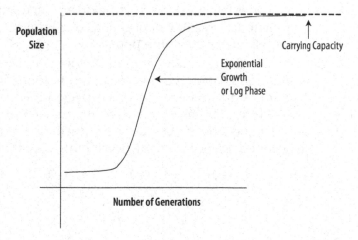

Be prepared to interpret population growth graphs and survivorship curves on the AP® Biology exam.

3. There are two types of factors that limit the growth of populations.
   i. *Density-independent factors*—affect population size regardless of density; most likely factors include weather or natural disasters.
   ii. *Density-dependent factors*—affect the population size based on the density of the population; most likely factors include food, predation, migration, or disease.
      ➤ Limited resources reduce population size in a density-dependent manner because competition for limited resources becomes more intense as the number of individuals using those resources increases.
      ➤ Poisoning due to accumulating waste materials becomes more likely, and affects more members of a population, as population density increases.
      ➤ Predation may be a density-dependent factor limiting a prey population if a predator increases its rate of predation when prey density is higher.
4. Small populations are more likely to become extinct than larger ones because inbreeding reduces the number, health, and genetic variability of offspring, or a local natural disaster could eliminate the entire population.
5. After remaining steady for most of human history, the human population has been increasing exponentially since the 1600s due to increasing life expectancy and greater ability to exploit resources. However, the growth rate has slowed since the 1960s due to a reduction in birth rate in developed, as well as many developing, countries.

Community Dynamics

I. Key Concepts

A. A community is a group of interacting populations of different species that live in the same geographic area.

B. Species richness, species diversity, and community stability are major characteristics of communities.

C. Species interactions and competition for resources are the bases for community relationships.

D. A succession of different communities occurs over time on newly created areas or in habitats destroyed by natural disasters or human activities.

II. Species Richness and Diversity

A. *Species Richness*—includes the number of different species in a community; *Species Diversity*—includes not only the number of each species, but the size of each population.

   1. Species richness increases as latitude decreases.
      i. Communities closest to the equator, such as those found in tropical rain forests, have the greatest number of species.
      ii. Three hypotheses may explain the greater number of species in lower latitudes.
         ➤ More available sunlight year-round promotes higher primary productivity—plant or phytoplankton

growth—resulting in a greater base level at the lowest trophic level of food chains.

➤ A more stable climate may lead to a greater number of niches available for exploitation.

➤ Tropical communities are older than those farther from the equator because they were not destroyed by recent ice ages.

Species Richness and Distance from Equator

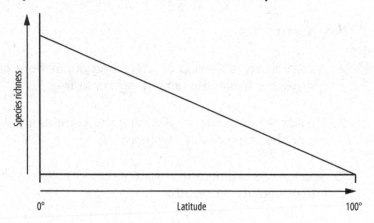

2. The *species-area effect* shows that species richness increases as the number of habitats and the area they cover increases. Therefore, larger islands have more species than smaller islands, and a reduction in habitat area is a main cause of the extinction of populations.

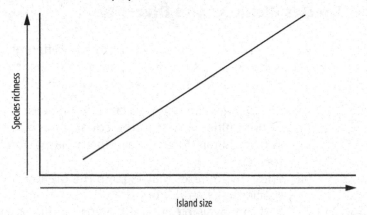

3. A *keystone predator* may increase species richness by preying on a successful competitor, thereby reducing competition between the prey species and its closest competitors and allowing those competitor populations to thrive.

Species Diversity with and without a Keystone Predator

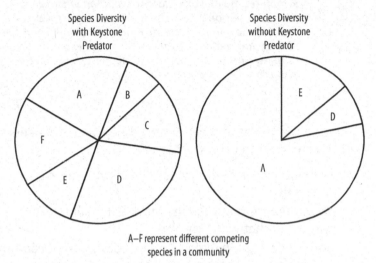

A–F represent different competing species in a community

B. *Communities with greater species richness are more stable* in the face of disturbances—such as droughts, floods, and other natural disasters—because a greater number of species tend to survive the disturbance in species-rich communities as opposed to species-poor communities.

III. Species Interactions

A. *Population Interactions*—interactions that occur with different species living in a community; can be beneficial to the species, but can also be detrimental to one or both of them.

B. *Symbioses*—close species relationships between a host and a symbiont

1. *Mutualism*—both species benefit from the relationship.

2. *Commensalism*—a relationship in which one species benefits while the other species is neither harmed nor helped.

3. *Parasitism*—beneficial to the parasite and detrimental—but not usually immediately fatal—to the host.

Test Tip

Symbioses may involve close coevolution where each species evolves in response to a change in the other species. Questions that require you to integrate material from different topics in biology are common on the AP® Biology exam, so be aware of these relationships as you study; for example, ecology and evolution are closely related.

C. *Predation*—predator eats the prey; one species benefits while obviously the other does not.

   1. *Adaptations for predator*—claws, teeth, poisons, speed, eyesight.

   2. *Adaptations for prey plants*—thorns in plants, plant chemicals that ward off prey.

   3. *Adaptations for animals*—cryptic coloration or camouflage, aposematic coloration or bright colors that warn, or warning noises.

   4. *Mimicry*—prey resembles another species.
      i. *Batesian mimicry*—a harmless species mimics a species that is dangerous to the predator.
      ii. *Müllerian mimicry*—two harmful species resemble each other and create a cumulative effect against a predator.

D. *Parasitism*—parasite lives off the host; one species benefits while the other does not (i.e., viruses, tapeworms, and mosquitoes).

E. *Competitive Exclusion Principle*—states that two species cannot survive in the same *ecological niche* (the sum of the total abiotic and biotic factors in an ecosystem) because they are competing for the same limited resources. Ultimately, neither species benefits from this interaction.

IV. Succession

A. *Succession*—the gradual progression of different communities over time that occurs on virgin territory or in a habitat recovering from natural or manmade disturbances and involves species changing the environment over time.

B. *Primary Succession*—begins slowly and takes longer than secondary succession (hundreds to thousands of years) because it occurs in areas that have not recently supported life.

　1. An area involved in primary succession may be newly exposed or newly formed rock, such as rock exposed by a glacier receding or islands produced by volcanic action.

　2. Autotrophic bacteria, algae, and lichens that grow on rocks are common pioneer species—the first species to colonize the area. Soil eventually will be produced to support plants and insects, and more gradually, larger animals and plants will migrate to the area and form a community.

C. *Secondary Succession*—occurs in areas where soil still remains and where communities used to exist but have been destroyed by disturbances such as fire, farming, and mining; this process can be completed within a year.

Test Tip

In succession, existing species change the environment over time, making it more favorable for other species to better adapt to the new conditions and outcompete them.

Ecosystem Dynamics

I

Key Concepts

A. An ecosystem includes groups of interacting communities and their environment; thus, it includes both *biotic* (living) and *abiotic* (nonliving) components.

B. A trophic level is an organism's nutritional position in a food chain.

C. Energy flows through an ecosystem in a one-way direction, from the sun to progressively higher trophic levels, passing along only about 10% of stored energy at each trophic level while traveling from lower to higher levels.

D. In contrast to a stream of energy running through an ecosystem, nutrients and water are recycled in an ecosystem.

E. Exponential growth of the human population has had effects on every aspect of the biosphere, from the global level to individual species.

II

Energy Flow

A. *Trophic Levels*—division of organisms in an ecosystem; energy flows through an ecosystem from lower trophic levels to higher trophic levels.

  1. *Primary Producers*—comprise the first trophic level and are almost exclusively dependent on solar energy; they are

photosynthetic organisms, such as plants and blue-green algae.

2. *Consumers*—organisms that comprise subsequent levels and that are ultimately dependent on producers for their energy needs.

    i. *Primary Consumers*— herbivores or plant-eating organisms.

    ii. *Secondary Consumers*—carnivores that eat the primary consumers.

    iii. *Tertiary Consumers*—carnivores that eat other carnivores or organisms below them.

3. *Detritivores*—derive their energy from dead organisms or detritus (i.e., fungi and soil microbes); they are extremely helpful in recycling matter.

B. *Food Chains and Food Webs*—illustrate energy interactions between members of a specific community.

1. *Food chains*—are single energy pathways in which food is transferred from one trophic level to the next.

2. *Food webs*—an elaborate web of organisms feeding at more than one trophic level.

Example of a Food Chain (----▸) within a Food Web (——▸)

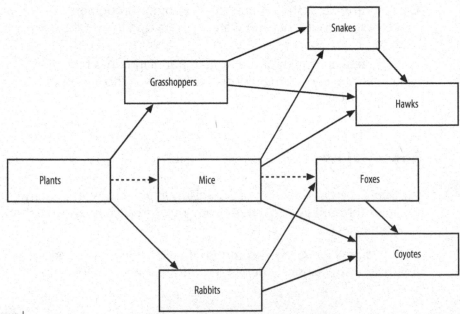

Test Tip

The arrows in a food chain or web indicate the flow of energy. If all arrows originating from a species point away from it, it is most likely a primary producer. If all arrows point only toward a species, it is either a consumer or a decomposer.

III. Nutrient Cycles

A. *Nutrient cycles* recycle water, carbon, nitrogen, and phosphorus, which move the ecosystem's organic matter to the abiotic portions of an ecosystem and back again.

B. *Water Cycle*—involves the processes of *evaporation, transpiration,* and *precipitation.*

C. *Carbon Cycle*—involves the processes of photosynthesis, cellular respiration, and combustion.

D. *Nitrogen Cycle*—involves a complex series of biochemical reactions by different soil bacteria to compounds assimilated by plants.

E. *Phosphorus Cycle*—provides the element phosphorus needed by all organisms as a component of nucleic acids and ATP.

Comparison of the Different Nutrient Cycles

Nutrient Cycle	Use in Organisms	Major Reserves	Important Processes
Water	Most of the mass of an organism is water	Bodies of water (oceans, lakes, rivers, streams)	Evaporation Transpiration Precipitation
Carbon	Organic molecules	Organisms Atmosphere	Photosynthesis Respiration Combustion

(continued)

Comparison of the Different Nutrient Cycles (*continued*)

Nutrient Cycle	Use in Organisms	Major Reserves	Important Processes
Nitrogen	Proteins Nucleic acids	Organisms Atmosphere	Nitrogen fixation Ammonification Nitrification Assimilation Denitrification
Phosphorus	Nucleic acids ATP	Rock, soil Organisms	Weathering Decomposition

IV. Human Impact on Ecosystems

A. *Growth of the Human Population*—in recent times, the human population has increased significantly and has profoundly altered the biosphere from the global to the local level.

1. Human population has only risen to great numbers very recently (since approximately 1650 CE).

2. Humans have needs similar to other large animals, but their numbers, worldwide distribution, and unique ability to extract and utilize resources have a greater impact on the Earth than most other organisms.

   i.   The use of fossil fuels to supply energy needed for industry and to heat and cool homes has resulted in a global increase in atmospheric carbon dioxide, which is correlated with a rise in global temperature and severe weather.

   ii. The disposal of biological, industrial, and household wastes has significantly altered nutrient cycling. These wastes have introduced either totally new, or otherwise rare, toxic molecules into the environment that have significant adverse effects on humans and other species.

B. *Global Effects on the Ecosystem*

1. *Biological Magnification*—toxic chemicals being increased in concentration from one trophic level to the next. Biomass from one level is created from a larger biomass from the trophic level before. Top-level consumers are mostly affected. The best-known example is the use of DDT pesticide.

2. *Ozone Layer*—the ozone layer absorbs harmful UV light. CFCs or chlorofluorocarbons in aerosol cans and refrigeration units destroy the ozone by reducing it to oxygen.

3. *Greenhouse Effect*—carbon dioxide emissions from the burning of fossil fuels act as a trap of solar heat in the atmosphere. Increases in carbon dioxide warm the air and accelerate the greenhouse effect. This is thought to be the major cause of global warming. Deforestation is also a major contributor to the greenhouse effect.

4. Despite these harmful effects on the ecosystem, some success has been achieved in reversing some of them.
   i.   Thinning of the protective ozone layer of the atmosphere has been addressed.
   ii. An acid precipitation reduction plan has been implemented in some countries, including the U.S.
   iii. Bans or reductions on toxins—such as DDT, PCBs, and mercury—have been achieved in some countries.
   iv. Climate change due to rising temperatures of the Earth has spurred development of models to predict the likely consequences and to provide information for making appropriate plans where possible.

Major Effects of Human Intervention on a Global Level

Problem	Cause	Global Consequence	Solution Attempted
Depletion of ozone layer surrounding the Earth	Use of CFCs	–More UV radiation –Greater cancer risk in humans –Greater likelihood of DNA damage in organisms	Phase out use of CFCs
Acid precipitation (pH below 5.6)	Sulfur and nitrogen oxides released from burning fossil fuels, especially coal	–Increase in aluminum toxicity in plants –Death of vulnerable organisms –Damage to entire ecosystems	Reduce sulfur and nitrogen oxides through "cap and trade" programs
Toxins, such as DDT, PCBs, and mercury	Industrial and household waste	–Poisoning and death of species in multiple ecosystems –Toxicity to humans	–Ban the use of the toxin –Clear industrial waste of toxin before release
Global climate change	Increased greenhouse gases, mainly from the burning of fossil fuels	–Polar ice melts, sea level rises –More and larger storms –Extinction of organisms sensitive to altered temperatures –Negative effects on agriculture	–Reduce greenhouse gases –Alternative energy sources –Plan for flooding, intense weather, and effects on agriculture

C. *Increased Rate of Species Extinction*—habitat destruction, overexploitation, and introduction of exotic species and diseases are the greatest threats to biodiversity and are currently contributing to an increase in the rate of species extinction.

1. *Biodiversity*—the degree of variation of species in a given area; the area can encompass a community or the entire biosphere.
2. Increasing *extinction rates* are currently decreasing biodiversity.
3. The major causes of decreasing biodiversity are over-exploitation, introduction of exotic species and diseases, and habitat destruction.
4. Reasons for preservation of biodiversity include utilitarian and nonutilitarian consideration.
5. *Conservation biology* focuses on maintaining biodiversity and includes strategies targeting all ecological levels.

D. *Sustainable Development*—limiting further damage to biodiversity and the environment involves sustainable development.

1. The goal of *sustainable development* is to manage the ecosystems of the biosphere in a way that supports the prosperity of human populations in the long term.
2. A continuing goal in this process is to study how ecological systems work to provide the best information for making decisions and how to manage and utilize the Earth's resources in a way that continues to replenish vital resources, such as clean air and drinking water, for future generations.

PART III

TEST-TAKING STRATEGIES AND PRACTICE QUESTIONS

Practice Multiple-Choice Questions

Practice with the following AP®-style questions. Then go online to access our timed, full-length practice exam at *www.rea.com/studycenter*.

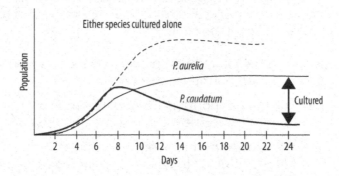

1. The graph above shows the population growth when two species of paramecia (*P. aurelia* and *P. caudatum*) are allowed to grow in separate flasks and together in one flask. Which of the following statements best explains the difference in growth?

    (A) *P. aurelia* produces toxic waste which harms both species, but harms *P. caudatum* more.

    (B) *P. aurelia* outcompetes *P. caudatum* for resources when they are grown in the same flask.

    (C) *P. aurelia* has a higher reproductive rate than *P. caudatum*.

    (D) When grown in the same flask, the two species reproduce, creating a hybrid and thus decreasing the number of both species.

The effects of the hormone known as growth hormone (GH) are mediated via the GH receptor (GHR). GHR is a 620-amino acid type I transmembrane protein that is expressed at high levels in liver and adipose (fat) tissue. The GHR extracellular region, where the hormone binds, contains seven cysteine residues, of which six are paired through disulfide bridges. GH binding causes the activation of the tyrosine kinase enzyme called Janus kinase 2 (JAK2) in the cell. Activated JAK2 phosphorylates the signal transducers as well as the activators of transcription molecules in the cell. These transcription molecules translocate to the nucleus to promote gene transcription.

2. Based on the information provided, which statement about growth hormone must be true?

(A) The growth hormone must be a nonpolar hormone because it can activate transcription.

(B) The growth hormone must be a polar hormone because it binds to an extracellular region of the receptor.

(C) The growth hormone must be a protein hormone because it has 620 amino acids.

(D) The growth hormone must be a steroid hormone because it binds to a transmembrane receptor.

Questions 3–4 are based on the following passage.

In deer mice, *Peromyscus maniculatus*, there are 2 tail lengths: long or short. A pair of long-tailed mice has 100 babies over their lifetime. The mice had 86 long-tailed babies and 14 short-tailed babies. Perform a chi-square test to determine whether the long-tailed trait is a single autosomal dominant trait.

3. What is the chi-square value?

(A) 6.5

(B) 3.2

(C) 10.1

(D) 0.59

4. Should the null hypothesis be rejected?

(A) Yes, because the *p*-value is greater than 0.05.

(B) No, because the *p*-value is less than 0.05.

(C) Yes, because the *p*-value is less than 0.05.

(D) No, because the *p*-value is greater than 0.05.

---

*Questions 5–6 are based on
the following material.*

A group of scientists created a plasmid (called the *SRC* plasmid) containing the following genes:

Sle: resistance to the antibiotic erythromycin

Ins: the gene for insulin

Cel: the gene that produces a binding protein that interacts with cellulose to turn on *Ins* gene

The scientists then used transformation techniques to insert the plasmid into some *E. coli* cells, while purposely not inserting the plasmid into others. They then plated the bacteria on several different types of agar plates. The results are shown below.

The following table shows the 5 plates that were set up and the data that were recorded after incubating at 37°C for 24 hours.

Plate	Was the plasmid inserted?	Is there erythromycin in the agar?	Is cellulose in the agar?
1	no	no	no
2	yes	no	no
3	yes	yes	no
4	no	yes	no
5	yes	yes	no

5. On which plate or plates would you **find live bacteria** that are **NOT** producing insulin.

(A) Plates 1 and 2 only

(B) Plates 1, 2, and 3 only

(C) Plates 1 and 4 only

(D) Plate 5 only

6. On which plate or plates would you find bacteria that contain the gene for insulin?

(A) Plates 2 and 3 only

(B) Plates 3 and 4 only

(C) Plates 2, 3, and 5 only

(D) Plate 5 only

7. In one type of fermentation, shown in the picture below, pyruvate is reduced to lactate. What is the purpose of this reduction?

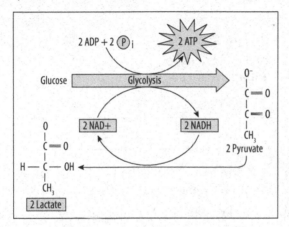

(A) Causes the creation of NAD+.

(B) It forms lactic acid needed for the electron transport chain.

(C) It generates ATP.

(D) It oxidizes pyruvate to make more ATP.

---

*Questions 8–9 are based on
the following material.*

In a tank, students kept an aquatic plant called Elodea, a fish known as a guppy, and a sea star. The mass of the organisms was measured, then the tanks were kept in the dark for 24 hours. Students in a class measured the mass of various living organisms. They then kept the organisms in the dark for 24 hours before remeasuring the mass. No nutrients were in the tank. The data are as follows.

Organism	Starting Mass	Final Mass
Elodea (submerged aquatic plant)	8.5 g	7.5 g
Sea star	20.2 g	19.83
Guppy	32.5 g	29.98 g

8. Which of the following is the best explanation for the pattern of change in mass of the organisms over time?

    (A) Water loss due to evaporation

    (B) Catabolism of stored glucose

    (C) Photosynthesis occurring

    (D) Growth and reproduction of the organisms

9. What is likely to happen to the oxygen level in the water in the tank while it is in the dark?

    (A) The oxygen level will go up due to the photosynthesis of the Elodea.

    (B) The oxygen level will stay the same, because dissolved oxygen does not diffuse out of water into organisms.

    (C) The oxygen level will go down due to cellular respiration of all three organisms.

    (D) The oxygen level will go down due to photosynthesis of the Elodea.

10. A patient is unable to use glucose for cellular respiration, but is able to use fatty acids and amino acids. Which of the following may be a reason for this condition?

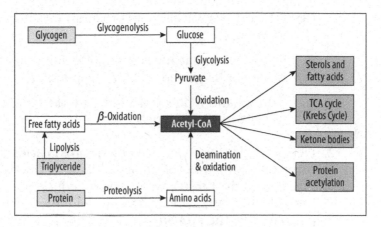

(A) His mitochondria lack the transport protein that moves pyruvate into the mitochondria.

(B) His cells cannot move NADH from glycolysis into the mitochondria.

(C) His cells contain something that inhibits ATP synthase in the mitochondrion.

(D) The enzymes that perform lipolysis are missing.

11. It is a misconception to think that the giraffe's neck became longer over time due to its neck stretching as generations of giraffes reached for ever higher leaves. Which statement explains why this is a misconception?

    (A) Overproduction of offspring leads to a struggle for survival and only the giraffes with longer necks survive to reproduce.

    (B) Not all traits will allow an organism to survive to reproduce.

    (C) If an organism acquires a trait after it is born, then it will not be passed on to its offspring.

    (D) If an organism does not use a part of its body, it will eventually disappear.

12. Pictured below are two of the many amino acids that make up a protein. Looking at just the side chains (the part in the boxes), what can be deduced about the polarity of the two amino acids and their resulting placement in the protein?

    Threonine   Alanine

    (A) Both are polar and hydrophilic and therefore will be on the inside of the protein structure.

    (B) Threonine is polar and hydrophilic, and will be on the outside of the protein structure, and alanine is nonpolar and hydrophobic, and will be on the inside of the protein structure.

    (C) Both are polar and hydrophilic and will be on the outside of the protein structure.

    (D) Threonine is polar and hydrophilic, and will be on the inside of the protein structure, and glycine is nonpolar and hydrophobic, and will be on the outside of the protein structure.

> *Questions 13–14 are based on
> the following material.*

The information below describes the dimensions of cuboidal cells in the epithelium (lining) of the human thyroid gland. The letters represent values that must be calculated.

	Cell 1	Cell 2
Side mm	5	A
Surface area mm²	B	486
Volume mm³	D	C
Surface Area-to-Volume Ratio	E	F

13. Which of the following fills in the table correctly for the dimensions of the cube?

    (A) A=7, B=150, C=294, D=125, E=6:5, F=2:7

    (B) A=9, B=150, C=729, D=125, E=6:5, F=2:3

    (C) A=9, B=50, C=486, D=150, E=8:5, F=3:2

    (D) A=8, B=50, C=294, D=175, E=4:3, F=2:6

14. Into which cell would oxygen diffuse more efficiently?

    (A) Cell 1 because it has a larger surface area-to-volume ratio.

    (B) Cell 2 because it has a larger volume, so there is more energy for diffusion.

    (C) Cell 2 because it has a larger surface area and thus contains more protein pumps for oxygen transport.

    (D) The oxygen diffusion efficiency is the same for both cells, since oxygen is such an important molecule.

Question 15 is based on
the diagram below.

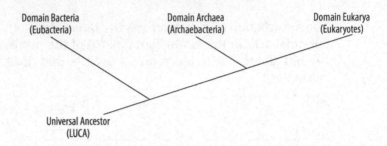

Domain Bacteria
(Eubacteria)

Domain Archaea
(Archaebacteria)

Domain Eukarya
(Eukaryotes)

Universal Ancestor
(LUCA)

15. The diagram above shows the relationship between the three domains. Which of the following characteristics must LUCA have had to be the universal common ancestor?

(A) Chloroplasts

(B) Mitochondria

(C) Ribosomes

(D) Nucleus

Nitrate salts are added to food as a preservative. Nitrite salts and amino groups react to form nitrosamines. Nitrosamines are potent mutagens. Since nitrates contain amino groups, there are concerns about using nitrates to preserve meats.

An experiment was performed to test the effect of adding vitamin C (ascorbate) on the mutagenesis of nitrates on bacteria living on meats. Bacterial cells were incubated with cured meat extracts in either the presence or absence of ascorbate. The rate of mutation for the bacteria cells was determined. The data are below.

Concentration of preserved meat extract (microgram/mL)	Concentration of ascorbate (microgram/mL)	Rate of mutation of bacteria cells (number of mutant bacteria per million cells)
0	0	2
0	50	2
10	0	8
10	50	2
20	0	18
20	50	2
30	0	40
30	50	2

16. Based on the data above, what is the most likely mechanism for ascorbate's effect on mutagenesis?

(A) Ascorbate likely causes mutations along with nitrosamines.

(B) Ascorbate likely inhibits an enzyme that converts nitrates to nitrosamines.

(C) Ascorbate likely enhances the interaction between nitrosamines and DNA.

(D) Ascorbate activates the genes that are mutated to be expressed.

17. Examine the figure below. If each arrow represents a division in mitosis, what can be said about the amount of genetic material in the G_1 phase in cell 1 compared to the G_1 phase in cell 5?

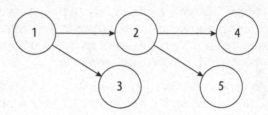

(A) Cell 1 has ¼ the amount of genetic material as cell 5.

(B) Cell 1 has ⅓ the amount of genetic material as cell 5.

(C) Cell 1 has the same amount of genetic material as cell 5.

(D) Cell 1 has twice the amount of genetic material as cell 5.

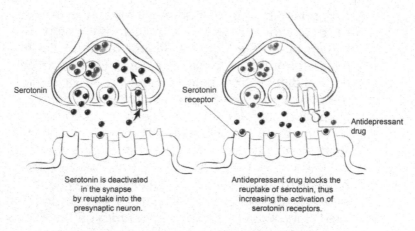

Serotonin

Serotonin receptor

Antidepressant drug

Serotonin is deactivated in the synapse by reuptake into the presynaptic neuron.

Antidepressant drug blocks the reuptake of serotonin, thus increasing the activation of serotonin receptors.

A: No Antidepressant Drug Present   **B: Antidepressant Drug Present**

18. Serotonin is a neurotransmitter that is released by a presynaptic neuron and binds to receptors on a postsynaptic neuron. When serotonin binds with ligand-gated sodium ion channels, it causes ligand-gated sodium ion channels on a postsynaptic neuron to open. Sodium ions then diffuse into the postsynaptic neuron, which results in a nerve impulse being sent. Serotonin is then taken up by the presynaptic neuron. Based on the figures above and the information provided, what is the result of taking the antidepressant drug?

    (A) The antidepressant drug will cause serotonin to be taken up by the presynaptic neuron more quickly.

    (B) The antidepressant drug will cause serotonin to remain in the synapse longer.

    (C) The antidepressant drug blocks the sodium ion channels and no nerve impulses can be sent.

    (D) The antidepressant drug causes more serotonin to be released from the presynaptic neuron.

Cortisol is a mammalian hormone released in response to stress. When released, cortisol causes the nervous system to cause a "fight" or "flight" response. Once released from the adrenal gland into the bloodstream, cortisol diffuses from blood plasma into saliva, enabling the collection of measurable amounts of hormones from saliva samples. The table below gives cortisol concentrations in saliva samples taken from two female Asian elephants over a six-day period during which the animals were introduced to each other at a zoo. The baseline (normal) salivary cortisol level in elephants is 4–5 nanomoles per liter. Elephant 1 had previously lived alone and was flown in from Africa on Day 1 prior to the introduction. The elephants were introduced to each other on Day 3.

Time (Days)	Elephant 1 salivary cortisol concentration in nmol/L	Elephant 2 salivary cortisol concentration in nmol/L
1	39	4
2	11	5
3	20	10
4	22	13
5	13	9
6	5	5
7	4	4

19. Which selective advantage would these changing cortisol levels give this species of elephants?

(A) When new elephants meet in the wild, they will be less likely to fight and more likely to cooperate with one another.

(B) When new elephants meet in the wild, the other elephants must be ready to defend their territory, or run if necessary.

(C) Since elephants live in herds, the reproductive hormones of the female elephants must be in sync with one another for more efficient reproduction.

(D) Elephants in the wild can release hormones that will affect the new elephants they meet.

Question 20 refers to the information below.

Isolated chloroplasts can perform photosynthesis without being within a plant cell. An experiment is done using chloroplasts that had been isolated from blended spinach leaves and stored in a cold, protective liquid. A colored dye is added to the liquid. This dye is blue when oxidized and colorless if reduced. The color change occurs only when chloroplasts are in the presence of light.

$$\text{Oxidized blue dye} \xrightarrow{\frac{\text{chloroplast}}{\text{light}}} \text{Reduced colorless dye}$$

$$\text{Oxidized blue dye} \xrightarrow{\text{light only}} \text{Oxidized blue dye}$$

20. Which of the following statements is true based upon the information above and your knowledge of photosynthesis?

(A) Glucose must be causing the blue dye to become reduced.

(B) The rise in pH due to carbon dioxide consumption must be causing the blue dye to become reduced.

(C) Electrons generated in the light reaction must be causing the dye to become reduced.

(D) The consumption of oxygen must be causing the blue dye to become reduced.

The following diagram is the basis of Question 21.
It shows a food web in a terrestrial ecosystem.

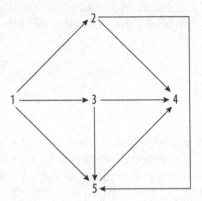

21. Which of the following would have the greatest impact on the energy contained in this ecosystem?

(A) Decrease the number of Species 4

(B) Increase the number of Species 5

(C) Increase the number of Species 1

(D) Decrease the number of Species 2

Aerobic yeast cells are placed in a solution of sugar. They are then attached to an apparatus that detects the number of bubbles of gas produced by the yeast cells. The rate of aerobic respiration varies with the surrounding temperatures as indicated by the data below.

Temp (degrees C)	5	15	25	35	45	55	65
Number of bubbles counted over 5 min.	0	15	35	60	35	15	0

22. Which gas forms the bubbles that are being counted?

    (A) Oxygen

    (B) Carbon dioxide

    (C) A mix of oxygen and carbon dioxide

    (D) Water

23. Which of the following is the best explanation for the data trend observed?

    (A) Enzymes are denatured at zero degrees Celsius.

    (B) Enzymes are denatured at temperatures above 45 degrees Celsius.

    (C) Carbon dioxide consumption is lowest at 35 degrees Celsius.

    (D) Oxygen diffuses out of cells best at 35 degrees Celsius.

Ling-Ling and Hsing-Hsing were two famous giant pandas that charmed visitors at the National Zoo in the 1980s. At that time, DNA evidence was just beginning to be used to analyze relationships between organisms. In the United States, the only other pandas in zoos were red pandas. However, these red pandas seemed more like raccoons to a lot of people, while giant pandas seemed more like bears. Scientists wanted to determine whether giant pandas and red pandas should belong to their own clade, or whether either is more closely related to bears or raccoons. They collected DNA from Ling-Ling and Hsing-Hsing along with brown bears, red pandas, and raccoons at the zoo. The table below shows part of the evidence collected.

The Number of Nucleotide Base Pair Differences in Cytochrome c Among Various Organisms

	Brown Bear	Giant Panda	Raccoon	Red Panda	Snake
Brown Bear	0	1	11	13	21
Giant Panda		0	10	12	20
Raccoon			0	3	18
Red Panda				0	17
Snake					0

24. Based on the data in the table, which organisms are represented by the letters in the cladogram below?

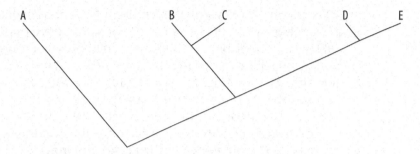

(A) A=snake, B=red panda, C=raccoon, D=giant panda, E=brown bear

(B) A=snake, B=giant panda, C=raccoon, D=red panda, E=brown bear

(C) A=snake, B=giant panda, C=brown bear, D=red panda, E=raccoon

(D) A=snake, B=brown bear, C=giant panda, D=raccoon, E=red panda

25. Which of the following is true?

(A) Red pandas and giant pandas share a more recent common ancestor than either does with the brown bear.

(B) Red pandas share a more recent common ancestor with raccoons than they do with giant pandas.

(C) Giant pandas and raccoons share a more recent common ancestor with each other than the giant panda shares with the brown bear.

(D) Snakes are the most recent common ancestor to all of the other organisms in the cladogram.

ANSWERS TO PRACTICE QUESTIONS

1. (B) is correct. If (A) were correct, then the P. aurelia would have the same growth curve whether P. caudatum were present or not. (B) is correct due to the competitive exclusion principle. No two species can occupy the same niche, as one will always outcompete the other. (C) is incorrect since the initial reproductive rate of P. caudatum is actually higher than that of P. aurelia. (D) is incorrect, since members of the same species cannot reproduce and make fertile offspring and this would not explain the difference in the number of individuals of each species when they are cultured together.

2. (B) is correct. (A) is incorrect because both hormone types (polar and nonpolar) can have the end result of activating transcription. (B) is correct since polar hormones cannot enter through the cell membrane and must have a receptor on the outside of the cell. (C) is incorrect because the 620 amino acids are actually part of the receptor. (D) is incorrect because steroid hormones enter cells and bind to a cytoplasmic receptor.

3. The correct answer is (A). Here is the chi-square formula. It will be given to you on the exam's formula sheet. o = observed data while e = expected data.

$$\chi_c^2 = \sum \frac{(o_i - e_i)^2}{e_i}$$

The way to calculate a chi-square is for each category, take the difference between observed number (which are the 86 long-tailed and 14 short-tailed mice) and the expected number (which is 75 and 25 if the trait were autosomal dominant). You then square each number and divide by the expected number for each category. The 2 numbers are then added together. So for the long-tailed mice, $86 - 75 = 11$. 11 squared is 121. 121 divided by $75 = 1.62$. For the short-tailed mice, $14 - 25 = -11$. -11 squared is 121. 121 divided by 25 is 4.84. Add 1.62 and 4.84.

4. (C) is correct. The null hypothesis states that there is no difference between the observed and the expected values. If the *p*-value is below 0.05, the null hypothesis is rejected. If the *p*-value is above 0.05, then the null hypothesis is not rejected. Another way to analyze the *p*-value is based on the critical value. The critical value is the chi-square value equivalent to 0.05. If the chi-square is above the critical value, the null is rejected. The following chart will be given to you on the exam. It shows the critical values for *p*-values of 0.05 and 0.10. Since the chi-square is above the critical value of 0.05, the null is rejected. When reading the chart, it is important to use the correct degrees of freedom. This is 1 subtracted from the number of categories, so in this case it is 1.

Chi-Square Table
Degrees of Freedom

p	1	2	3	4	5	6	7	8
0.05	3.84	5.99	7.82	9.49	11.07	12.59	14.07	15.51
0.01	6.64	9.32	11.34	13.28	15.09	16.81	18.48	20.09

At 1 degree of freedom, the critical value for a *p*-value of 0.05 is 3.84. The calculated chi-square is above that, so the null hypothesis should be rejected. The *p*-value is less than 0.05.

5. The correct answer is (B). All bacteria will grow if there is no antibiotic present (erythromycin). However, they would need the plasmid and the cellulose to produce insulin. If erythromycin is present, the plasmid would be necessary for them to grow. So, bacteria will grow on plates 1, 2, and 3. On plate 4, erythromycin will kill the bacteria and on plate 5, bacteria will grow, but will also produce insulin.

6. (C) is correct. Any of the plates with bacteria that received the plasmid will contain the insulin gene, even if it is not expressed. So plates 2, 3, and 5 will all have the insulin gene.

7. (A) is correct. During glycolysis, NAD+ acts as an oxidizing agent in order to oxidize glucose to pyruvate. When NAD+ is reduced, it becomes NADH. In aerobic respiration, this NADH gets oxidized during the electron transport chain. NAD+ is therefore formed again and can be reused in glycolysis. Since in fermentation, there is no electron transport chain, the NADH needs an alternative pathway to be replenished. As shown in the diagram, NAD+ is formed during the conversion of pyruvate to lactic acid, and that NAD+ enters glycolysis again. (B) is incorrect since there is no electron transport chain in anaerobic respiration. (C) and (D) are both incorrect, as no formation of ATP is shown in the diagram.

8. The correct answer is (B). (A) is incorrect, as the organisms are submerged in water; thus, no water will evaporate from them. (B) is correct, because at all times, organisms must be catabolizing glucose or other organic molecules in order to transfer energy to ATP. That process produces carbon dioxide, which leaves the organism as a gas, thus decreasing the organism's mass. (C) is incorrect. Photosynthesis adds to the mass of organisms, but the mass of the organisms decreased. (D) is incorrect, since the mass would increase if the organisms could grow. The organisms cannot grow since there were no nutrients added to the tank.

9. (C) is correct. (A) is incorrect, as the Elodea cannot do photosynthesis in the dark. (B) is incorrect because the organisms need oxygen from the water to perform photosynthesis. Choice (C) is correct since the organisms need oxygen to perform cellular respiration and create ATP. (D) is incorrect because photosynthesis, if it were performed, produces more oxygen. However, it is not performed since the plants are in the dark.

10. The correct answer is (A). The model shows that glucose is first converted to pyruvate before it can become acetyl-CoA and enter the Krebs Cycle (TCA cycle). It is known that the TCA cycle occurs in the mitochondrion. (B) is incorrect, as there is also NADH produced during the TCA cycle which can be used for ATP production, so this would not prohibit the

energy from glucose being used. (C) is incorrect, as ATP synthase would be needed in the metabolism of amino acids and
fatty acids as well. All three types of monomers get converted
into acetyl-CoA, after which the process is the same for all. (D)
is incorrect, since the diagram shows lipolysis is a process in
the metabolism of fatty acids, so absence of the enzymes that
perform it would not affect glucose metabolism.

11. (C) is correct. While (A) and (B) are true statements, they do
    not serve to correct the misconception. Both (A) and (B) can
    be correct with regard to the statement. (C) makes the state-
    ment impossible. Organisms cannot pass on traits that they
    gain during their lifetime, so the stretched necks cannot be
    passed to offspring. (D) is incorrect and would not correct the
    statement even if it were true.

12. The correct answer is (B). (A) is incorrect because while side
    chain of threonine has an OH⁻ group that makes it hydro-
    philic, alanine's side chain only has a methyl group, which is
    hydrophobic. Choice (B) states the polarity correctly. Because
    threonine is hydrophilic, it will be attracted to water in the
    environment of the protein and thus toward the outside of the
    final protein. Hydrophobic side chains like alanine are repelled
    by water, causing those side chains to be on the inside of the
    final protein. (C) is incorrect because the polarity is not stated
    correctly. (D) is incorrect because the position of the amino
    acid side chains in the final protein structure is incorrect.

13. (B) is correct. The formulae for surface area and volume are
    on the formula sheet that is given with the exam. Volume of
    a cube is calculated by cubing the length of any side. Surface
    area of a cube is calculated by squaring one side and multiply-
    ing that number by 6.

14. The correct answer is (A). The efficiency of diffusion is best
    for cells that have a higher surface area-to-volume ratio. (B) is
    incorrect because while the surface area is larger, there is more
    volume for substances to diffuse through. (C) is incorrect since
    no protein pumps are involved in diffusion. (D) is incorrect

since while all human cells need oxygen, the diffusion efficiency is dependent on physical size.

15. (C) is correct. LUCA is the common ancestor of all three domains. Thus, LUCA must have characteristics of all three domains. The only cell part listed that is in each of the domains is ribosomes.

16. (B) is correct. From the data it is clear that the bacteria on the meat with ascorbate have fewer mutations than the meat without ascorbate. (A) is incorrect since it says that ascorbate causes mutations. Choice (B) would result in fewer mutations, since mutations are caused by nitrosamines, and inhibiting the enzyme that causes production of nitrosamines would reduce the amount of this mutagen. (C) is incorrect because increasing the binding between the mutagen and DNA would increase mutations. (D) is incorrect because it results in more gene expression rather than more or fewer mutations.

17. The correct answer is (C). Mitosis conserves the amount of DNA with each division due to DNA replication happening prior to mitosis. This creates two identical sister chromatids that separate into each of the new cells.

18. (B) is correct. (A) is incorrect because the model shows the antidepressant drug blocking the protein in the presynaptic membrane that allows for the reuptake of serotonin. (B) is correct, since by blocking the protein in the presynaptic membrane that allows for the reuptake of serotonin, the serotonin will stay in the synapse longer. (C) is incorrect because the sodium ion channels are in the postsynaptic membrane and serotonin interacts with the presynaptic neuron. (D) is incorrect, since no interaction between the vesicles that release neurotransmitters and the antidepressant drug occurs.

19. The correct answer is (B). The data shows cortisol increasing when the elephants meet. The paragraph states that this leads to preparation for "fight or flight." (A) is incorrect since it indicates that the elephants would be calmer when meeting. (B) is correct, as cortisol increase would help the elephants deal with

a stressor. (C) is incorrect because cortisol is not described as a reproductive hormone. (D) is incorrect, as hormones are released and have their effect within an organism. If cortisol could affect other elephants, it would induce fight or flight in those elephants and help the other elephants succeed in a fight if one were to occur. This would not help the elephant releasing the cortisol to survive to reproduce.

20. (C) is correct. (A) is incorrect since glucose is not oxidized in chloroplasts. The oxidation of glucose occurs in mitochondria and partially in the cytoplasm. (B) is incorrect. While pH does rise when carbon dioxide is used, pH change does not cause reduction. The correct answer is (C) because reduction occurs when a molecule gains electrons. Free electrons are released during the light reaction and they can then reduce the dye. (D) is incorrect since oxygen is not consumed by chloroplasts.

21. (C) is correct. Energy contained in any ecosystem originates in the producers. Thus, the only way to impact the total energy is to increase or decrease the producers. Increasing Species 1 will mean there is more photosynthesis and therefore more energy available for the other trophic levels. On this diagram, there are arrows from Species 1 to Organisms 2, 3, and 4. No arrows point to Species 1.

22. (B) is correct. Since these yeast cells are aerobic, they produce carbon dioxide. Yeast are not photosynthetic, so no oxygen is being produced.

23. The correct answer is (B). (A) is incorrect since low temperatures do not cause the denaturing of enzymes, but the low temperatures do slow the molecules and cause fewer collisions per minute. (B) is correct since there is little to no enzyme action above 45 degrees Celsius and heat will cause vibrations which cause the intermolecular forces in the secondary and tertiary structures of the proteins to be disrupted, thus denaturing the enzyme.

24. (A) is correct. The branch that is farthest from the other branches is the snake, because the snake is put into the analysis as an outlier. An outlier is an organism that is known to have a less recent common ancestor than any of the others and is used as a control to be sure that the analysis makes sense. If the outlier were to appear to share a more recent common ancestor with the other organisms, then it is known that the method is faulty. Because there are the fewest differences between the raccoon and red panda, and between the brown bear and giant panda, it is known that each pair belongs together on a branch, showing that they share a recent common ancestor. The raccoon and red panda should be closer to the snake, since they have slightly fewer differences than the brown bear and the giant panda.

25. (B) is correct. The amino acid differences are lowest between organisms that share the most recent common ancestors. The raccoon and red panda only differ by 3 amino acids and the brown bear and giant panda only differ by 1 amino acid.

Science Practices and Essay Writing

I. General Tips on Essay Writing

A. *Overall Concept*—The AP® Biology course is unified by thematic underpinnings that can be related to any of the topics presented in this *AP® Biology Crash Course.* A good way to utilize these themes is to incorporate them into your essays in the free-response section of the exam. The AP® Biology readers will be delighted that you were able to see the big picture of the course rather than small, isolated concepts. Below are the major themes:

Big Idea 1—The process of evolution drives the diversity and unity of life.

Big Idea 2—Biological systems use energy and molecular building blocks to grow, reproduce, and maintain dynamic homeostasis.

Big Idea 3—Living systems store, retrieve, transmit, and respond to information essential to life processes.

Big Idea 4—Biological systems interact, and these systems and their interactions possess complex properties.

B. Tips for Essay Writing—Consider the following important factors when writing an essay for this exam:

1. You are not being graded on the style or format of your answer but rather on your ability to communicate in every sentence the science they are looking for. Do not write an introduction, conclusion, or a thesis statement. That just wastes precious time that should be spent responding to the actual substance of the question.

2. Make sure your sentences address the task verbs in the prompt. These verbs are bolded. Below is a description of each verb and what it means:

These **task verbs*** are commonly found in the free-response questions:

Calculate: Perform mathematical functions and determine a numerical answer with units.

Construct/Draw: Explain a concept or experimental design using an illustration. Most likely you'll need to label the illustration.

Describe: Elaborate on a topic using characteristics of the process, substance, or organism.

Determine: State a conclusion or decision after examining the evidence given.

Evaluate: Determine whether a claim is valid or whether information is significant.

Explain how: Analyze the relationship, pattern, position, situation, or outcome.

Explain why: Analyze the reasons for a relationship, pattern, position, situation, or outcome.

Identify: State, but do not elaborate, the process, organism, or substance indicated in the question.

Justify: Give evidence to support a claim.

Make a claim: From the evidence given, make an assertion about what is happening or what will happen.

Predict/Make a prediction: State what you think is going to happen.

Represent: Describe biological relationships using graphs or illustrations.

State (the null/alternative hypothesis): The null hypothesis means that there is no difference between groups, while the alternative hypothesis predicts that one is higher than the other. These are statistical hypotheses and do not necessarily indicate what you predict will occur.

Support a claim: Explain how evidence given or calculated supports a claim.

*   Adapted from College Board AP® Biology Course and Exam Description, © 2020, p. 218.

3. Look for biology terms in the prompt. These terms are important to determine the answer. For example, if two "species" are mentioned, likely your knowledge of what a species is, is what is being sought. Other common terms to look for are "energy," "photosynthesis," "ATP," "structure and function," "prokaryotic," "eukaryotic," and "molecular level."

4. Reread your answers if you have time. Make sure your responses address each task verb and demonstrate a clear understanding of any biology terms in the prompt.

II. Must-Know Science Practices

The following is a list of the six science practices that the AP® Biology exam will assess. Keep this list in mind as you round out your preparation.

A. Science Practice 1—Concept Explanation

1. Use or create models to describe and explain biological concepts and processes.
2. Models may be mathematical, diagrams, or flow charts.
3. If a model is given, be able to interpret and explain the model.

B. Science Practice 2—Visual Representations (Examples include Mathematical Models, Diagrams, and Flow Charts)

1. Use or create proper graphs and data tables to represent concepts, processes, or models.
2. Explain relationships between different concepts and processes and how these representations relate to larger biological principles.

C. Science Practice 3—Questions and Methods

1. Design an appropriate experiment or evaluate an experimental design.
2. Write a testable question, state a hypothesis, determine appropriate control and experimental groups, identify independent and dependent variables, collect appropriate

data, evaluate the data, and decide on the next experiment that should be done.

D. **Science Practice 4—Representing and Describing Data** (Examples include Bar; Line; Box and Whisker; Pie; Histogram; Log Y; and Dual Y)
   1. Create or describe data in a table, graph, or other appropriate figure.
   2. Account for proper orientation, labeling, units, scaling, plotting, type of figure.
   3. Plot points, identify trends or patterns or relationships between variables, as necessary.

E. **Science Practice 5—Statistical Tests and Data Analysis**
   1. Use mathematics appropriately.
   2. You may be asked to calculate a rate, mean, ratio, percentage, or chi-square value.
   3. You must also be able to determine what the answers to these calculations mean.
   4. You will be given a formula sheet and may be asked to use any formulas on that sheet.

F. **Science Practice 6—Argumentation**
   1. Claim-Evidence-Reasoning.
   2. State a scientific claim based on scientific principles, provide evidence from your knowledge or the prompt to substantiate that claim, and explain how the evidence relates to the claim.
   3. You should also be able to make and justify predictions (cause and effect).

PART IV

AP® BIOLOGY LABS

The AP® Biology Labs

I. The Lab Mindset

There are no specific labs you need to know for the AP® Biology exam. However, you will encounter exam questions that require you to apply science practices to real-world scenarios. You will thus need to know how to design and conduct valid experiments. As a result, understanding the six science practices covered in Chapter 26: Science Practices and Essay Writing is pivotal to your success on the exam.

Be ready to use your applied knowledge rather than relying on simple factual recall. You may see unfamiliar lab material on the exam: Don't be flustered by it. Focus instead on using the science practices "to develop and refine testable explanations and predictions of natural phenomena," as the College Board puts it in the latest AP® Biology Course and Exam Description.

Test Tip

There are no required labs that will be tested on the exam, but if you understand the labs discussed here, you will be able to answer questions that focus on interpreting and evaluating experimental results.

Though labs literally are not required on the AP® Biology exam, you should expect the six science practices that define and frame the lab experience to show up in exam questions. The exam requires you to be able to reason your way through a scientific argument, including devising a plan for collecting data, analyzing the data, using appropriate mathematical calculations, and substantiating your findings with evidence.

LAB 1   Artificial Selection

Analysis Question: How will you know if artificial selection has changed the genetic makeup of your population of plants?

Exercise 1A: Analyzing Plant Trichomes (1st Generation)

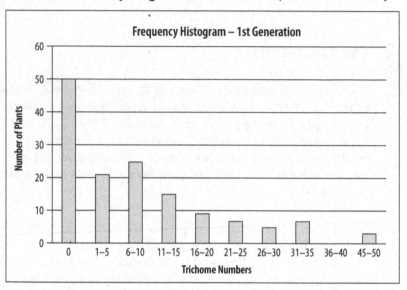

Interpretation of this Exercise

➤ This is an exercise in understanding the characteristics of natural selection and artificial selection.

➤ This first histogram shows a selection of plants with the characteristic trichome, "plant hairs."

➤ You may be asked to determine if this figure is representative of natural selection or artificial selection. Specifically, this is an example of a random assortment of plants screened for a specific trait.

Exercise 1B: Analyzing Plant Trichomes (2nd Generation)

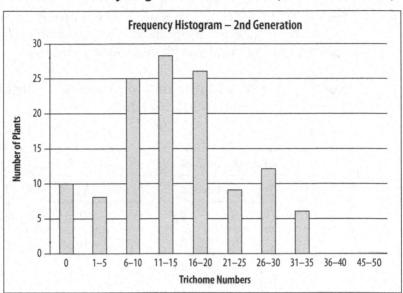

Interpretation of this Exercise

➤ This second figure demonstrates a second generation of plants that have been preselected to be grown based on having a desired characteristic (in this case, plant hairs).

➤ The goal of artificial selection is to favor a desired trait for a specific reason and is faster than natural selection at having organisms demonstrate this characteristic because the next generation can be absolutely restricted to offspring of parents that meet the desired criteria.

➤ In contrast, natural selection depends on the environment to do the selecting; therefore, traits like trichome would show up at a significantly slower rate than seen in two generations of plants.

➤ Questions on these figures may center on asking whether this figure demonstrates artificial selection and why. How do you know for sure? The answer is: Yes, this is an example of artificial selection, and this is evident because the number of plants with the desired trait significantly increased only within one generation. That would not be the case if natural selection were the only selecting factor.

LAB 2   Mathematical Modeling: Hardy-Weinberg

Analysis Question: How can the use of mathematical models be used to investigate the relationship between allele frequencies in populations of organisms with evolutionary change?

Exercise 2A: Using Mathematical Modeling to Test for Hardy-Weinberg Equilibrium

p (the frequency of A) = 0.26						
q (the frequency of B) = 0.74					Tasters	Nontasters
		Gametes		Zygotes	AA	AB	BB
		B	A	BA	0	1	0
		B	A	BA	0	1	0
		B	A	BA	0	1	0
		A	A	AA	1	0	1
		A	B	BA	0	1	0
		A	A	AA	1	0	0
		B	B	BB	0	0	1
		B	B	BB	0	0	1
		A	B	AB	0	1	0
Genotype					2	5	3
					9 A's		11 B's

Interpretation of this Exercise

➤ The spreadsheet above represents one version of mathematical modeling that examines all the gametes, zygotes, and specific alleles within a random sampling of the population. The spreadsheet calculated the frequency of p (A) and q (B). The main ideas are that natural selection, as part of evolution, can act on a phenotype and create variations within a population.

Evolutionary change is also driven by random processes, and populations of organisms continue to evolve.

Exercise 2B: Case Studies

CASE 1—A Test of Ideal Hardy-Weinberg

Interpretation of this Exercise

➤ A population that is in an Ideal Hardy-Weinberg would be a population of heterozygote individuals that follow all 5 key Hardy-Weinberg criteria:
  • No mutation
  • No gene flow or genetic variation
  • A very large population sample
  • No natural selection
  • Random mating

    Frequency of p and $q = 0.5$

    Percent of $p^2 = 25\%$

    Percent of $2pq = 50\%$

    Percent of $q^2 = 25\%$

CASE 2—Comparison to Hardy-Weinberg

Interpretation of this Exercise

➤ If this population is random with no known restrictions on the population, can this population be in Hardy-Weinberg equilibrium?
  • Frequency of $p = 0.26$ and $q = 0.74$
  • Percent of $p^2 = 10\%$
  • Percent of $2pq = 35\%$
  • Percent of $q^2 = 55\%$

➤ Yes, it's possible for this population to be in Hardy-Weinberg equilibrium, even though the above calculations don't support this conclusion. The reason is that the population sampled is too small to reflect accurately whether there is Hardy-Weinberg equilibrium.

Exercise 2C: Uses of Mathematical Modeling for Hardy-Weinberg

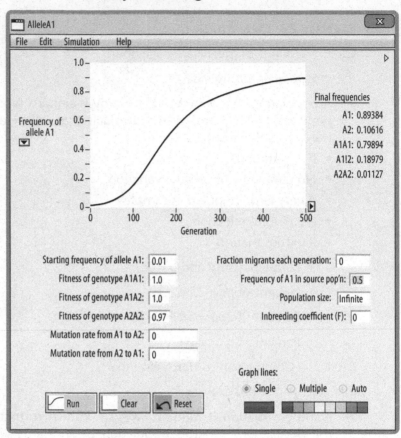

Interpretation of this Exercise

➤ The above figure shows an example of a program that can mathematically model the frequency of a specific allele by taking into account the starting frequency of that allele in a population and also the fitness of each of the genotypes. Once these numbers are applied to the program, then the program charts the increased frequency of this allele up to 500 generations.

➤ You may be asked to interpret the line chart on the previous page, which indicates that the frequency of allele A1 steadily increased its presence in the population over 500 generations, and at the 500th generation, its frequency in the population would be approximately 0.90.

➤ You also may be asked to explain the uses of this type of modeling. With the increased presence of computer programs able to articulate large amounts of data, programs such as the one above can quickly project the presence of specific alleles and take into account multiple factors that would alter its presence in one or more populations. These programs are useful learning tools for students to be able to plug in different allele frequencies and see the consequences on the chart; these programs are useful to researchers because they save time that otherwise would be spent calculating these projections by hand.

LAB 3 Comparing DNA Sequences to Understand Evolutionary Relationships with BLAST

Analysis Question: How can bioinformatics be used to help us better understand evolutionary connections and also the presence of genes in multiple organisms?

Exercise 3A: Constructing a Cladogram from Data in a Chart

Organism	Vascular Tissue	Seeds	Flowers
#1	0	0	0
#2	1	0	1
#3	1	1	1
#4	1	0	0
Total	3	1	2

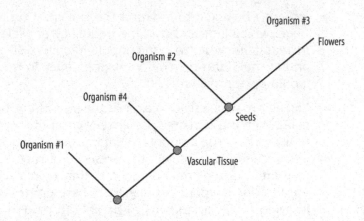

Interpretation of this Exercise

➤ Cladograms are used to map out the evolutionary connections between different organisms and also to demonstrate
which specific traits (e.g., vascular tissue, seeds, etc.) seem
to have evolved first in evolutionary history. This exercise is
meant to test your ability to articulate the data in the chart
and to draw the above cladogram. You also may be asked to
analyze a cladogram, like the one above.

Exercise 3B: Constructing a Cladogram from Gene Percent Similarities to Humans

Organism	Gene Similarity
#1	97%
#2	92%
#3	78%
#4	62%

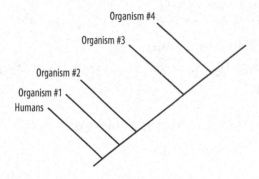

Interpretation of this Exercise

➤ The above chart shows the percentage similarity that each of the organisms #1–#4 are similar to humans. According to the above chart, organism #1 is the most similar, in reference to this gene, to humans. This exercise is meant to test your ability to articulate the data in the chart and to draw the above cladogram. This particular type of cladogram is meant to show the percentage similarities; therefore, the approximate placement of each of the branches is important.

➤ The BLAST sequence analysis tool can also be used to determine how similar a specific gene, nucleotide, or protein is to other organisms, which includes humans, mice, etc. The database is searched for the specified sequence and the results are listed by their percentage similarity.

➤ You may be asked to interpret data from a BLAST search that indicates the percentage similarities, just like the above chart. The skill is still the same: the higher the percentage similarity, the closer they should be mapped on the cladogram.

BIG IDEA 2: ENERGETICS

LAB 4 Diffusion and Osmosis

Analysis Question: What causes plants to wilt if you forget to water them?

Exercise 4A: Watering Plants and Turgor Pressure

Interpretation of this Exercise

➤ Normally, when a plant has enough water supply, the cell has enough water to "feed" itself as well as enough water to store in its vacuoles. These vacuoles, full of water, are what help the leaves stay more stiff by creating turgor pressure. So when a plant does not get watered for quite a while and begins to wilt, the plant has begun drawing water out of the plant's system, but more importantly, out of its vacuoles containing water.

Exercise 4B: Diffusion

	Initial Contents	Solution Color		Presence of Glucose	
		Initial	Final	Initial	Final
Bag	15% glucose and 1% starch	Clear	Blue/black	Yes	Yes
Beaker	H_2O & IKI	Yellowish	Yellowish	No	Yes

Interpretation of this Exercise

➤ Starch is too large a molecule to escape through the pores of the dialysis bag. As a result, the content of the dialysis bag turns from colorless to blue/black because the IKI from the beaker diffuses through the pores and reacts with the starch (a positive test for starch). The beaker fluid stays yellowish because no starch has diffused from the dialysis bag. Glucose is present in the beaker because it is a smaller molecule than starch and diffuses through the dialysis bag's pore. Residual glucose is still in the bag; thus, you continue to have a positive result for the bag.

Exercise 4C: Osmosis

Contents in Bag	Percentage Change
0.0 M Distilled Water	0.1%
0.2 M Sucrose	2.7%
0.4 M Sucrose	5.0%
0.6 M Sucrose	8.1%
0.8 M Sucrose	11.0%
1.0 M Sucrose	14.1%

% Change in Mass vs. Concentration of Sucrose

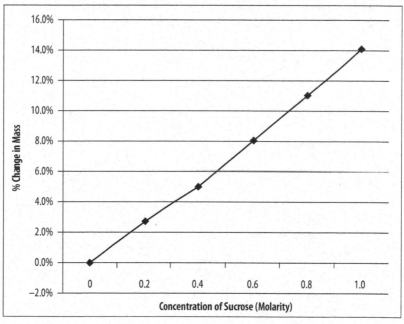

Interpretation of this Exercise

➤ As the concentration of the solute increases, water diffuses into the dialysis bag (hypotonic to hypertonic), increasing the mass of the dialysis bag. As an isotonic solution is evident with distilled water, diffusion is occurring at equal rates into and out of the dialysis bag.

Exercises 4D: Water Potential

Contents in Beaker	Percentage Change
0.0 M Distilled Water	19.0%
0.2 M Sucrose	8.0%
0.4 M Sucrose	–5.0%
0.6 M Sucrose	–13.0%
0.8 M Sucrose	–21.0%
1.0 M Sucrose	–27.0%

% Change in Mass vs. Concentration of Sucrose

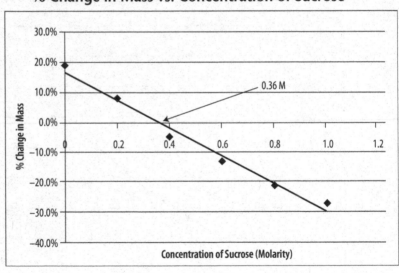

Interpretation of this Exercise

➤ The contents in the beaker and the percentage change in mass, as recorded in the chart above, is then graphed to demonstrate the relationship of the change in mass to the concentration of sucrose. Essentially, when the line crosses the *x*-axis at 0.36 M (estimation), then the concentration of the potato core is isotonic to the sucrose concentration. A net change of 0% at *x* = 0.36 M is the concentration that the water potential in the potato tissue is equal to the sucrose concentration (isotonic).

➤ Calculation of Water Potential

$$\psi_s = -iCRT = -(1)\left(.36\frac{\text{moles}}{\text{liter}}H_2O\right)\left(0.0831\frac{\text{literbar}}{\text{mole°K}}\right)(295°K) = -8.8 \text{ bars}$$

If the calculated water potential is less than the water potential surrounding the bag, then water will flow into the bag (more solutes/molecules inside the bag). If the calculated water potential is greater than the water potential surrounding the bag, water will flow out of the bag (less solute molecules inside the bag). Thus, water will flow from high to low water potential.

Exercise 4E: Plant Cell Plasmolysis

Interpretation of this Exercise

➤ Plant cells that are in a hypotonic solution will cause water to diffuse into the cell, thus creating a turgid cell.

➤ Plant cells that are in a hypertonic solution will cause water to diffuse out of the cell, thus creating a plasmolyzed cell.

➤ Plant cells that are in an isotonic solution will cause water to equally diffuse across the cell, making the cell flaccid.

LAB 5 Photosynthesis

Analysis Question: What factors affect the rate of photosynthesis in living leaves?

Exercise 5A: Median Rate of Photosynthesis

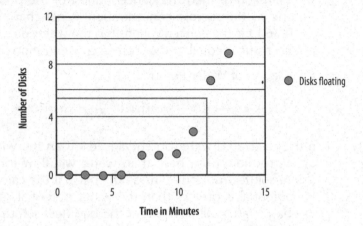

Interpretation of this Exercise

➤ Photosynthesis can be measured either by the production of O_2 or by the consumption of CO_2. The above chart shows how long it took each of the leaf disks to float to the surface—indicating that photosynthesis was taking place. The median of the disks floating is charted above, suggesting that it took approximately 12 minutes for half of the leaves to float to the surface.

Exercise 5B: Photosynthesis vs. Light Intensity

Light Intensity	Rate of Photosynthesis
0	0
200	0.05
400	0.06
600	0.07
800	0.08
1000	1.0
1200	1.1
1400	1.2

Interpretation of this Exercise

➤ The above chart shows that the increase of light intensity also increases the rate of photosynthesis. Be prepared to analyze this type of data either in chart form (as seen above) or on a line graph.

LAB 6 Cellular Respiration

Analysis Question: What factors affect the rate of cellular respiration in multicellular organisms?

Exercise 6A: How Temperature Affects Oxygen Consumption

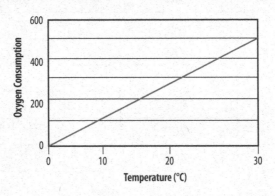

Interpretation of this Exercise

➤ The above line graph demonstrates that increased temperature is proportional to increased oxygen consumption. Be prepared to analyze data like that above and to compare the rates of cellular respiration for multiple organisms on the same graph.

BIG IDEA 3: INFORMATION STORAGE AND TRANSMISSION

LAB 7 Cell Division: Mitosis and Meiosis

Analysis Question: How do eukaryotic cells undergo mitosis or meiosis?

Exercise 7A: Observing Mitosis in Plant and Animal Cells Using Prepared Slides of Onion Root Tip and Whitefish Blastula

Interpretation of this Exercise

➤ Be able to draw a cell in Interphase (non-dividing portion of the cell cycle) and the 4 stages of mitosis.

Interphase

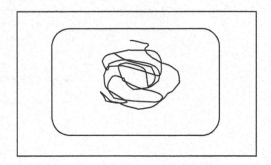

Prophase

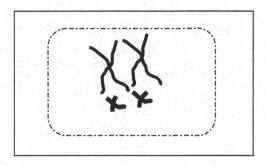

Metaphase

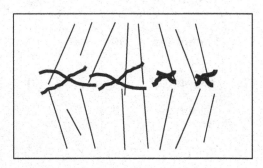

Anaphase

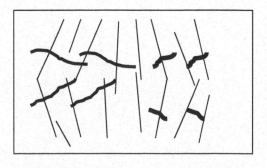

Telophase

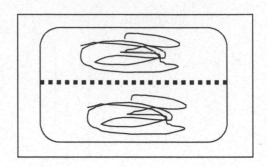

Exercise 7B: Time for Cell Replication

	Number of Cells				Percentage of Total Cells Counted	Time in Each Stage
	Field 1	Field 2 '	Field 3	Total		
Interphase	500	600	700	1800	82.0%	19 hr 41 min
Prophase	50	60	70	180	8.2%	1 hr 58 min
Metaphase	30	40	50	120	5.5%	1 hr 30 min
Anaphase	15	20	30	65	3.0%	43 min
Telophase	8	10	12	30	1.3%	19 min
				2195		

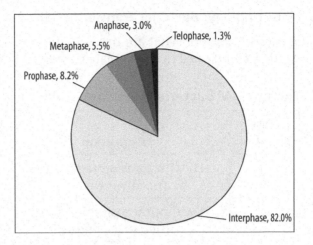

Interpretation of this Exercise

➤ The length of the cell cycle is roughly 24 hours with the majority of that time spent in Interphase, getting prepared for mitosis. Of the cells that one would observe in mitosis, the predominant phase is prophase.

Exercise 7C: Cancer and Mitosis

Interpretation of this Exercise

➤ Be prepared to answer questions about how cancer can affect the cell cycle:

- Increases the rate of mitosis.
- Cells spent less time "checking" whether everything is in order before continuing through the rest of the cell cycle; this "rush" causes mistakes to be made during replication and further mutations to daughter cells.
- Cells eventually can become so mutated that they do not resemble the original cells.
- Signals to indicate that the cell has not replicated properly and that the cell should undergo apoptosis (cell death) can be disregarded.

LAB 8 Biotechnology: Bacterial Transformation

Analysis Question: What are the ways we can utilize genetic engineering techniques to manipulate heritable information?

Exercise 8A: Bacterial Transformation

Plate Number	Condition	Observation
1	LB with transformed plasmid (positive control)	Lawn
2	LB without transformed plasmid (negative control)	Lawn
3	LB/Amp with transformed plasmid (experimental)	50 colonies
4	LB/Amp without transformed plasmid (positive control)	None

Interpretation of this Exercise

➤ Plate numbers 1 and 2 will have lawns (growth) of bacteria because there was no antibiotic in the plate agar.

➤ Plate number 3 had 50 transformed colonies because some of the cells were transformed with the plasmids containing the gene for resistance to ampicillin.

➤ Plate number 4 has no colonies since no plasmid was transformed, and the bacteria are susceptible to ampicillin.

Exercise 8B: Transformation Efficiency

$$\text{Total Efficiency} = \frac{\text{Total \# of colonies grown on agar plate}}{\text{Amount of DNA spread over agar plate}}$$

Interpretation of this Exercise

➤ Total mass of plasmid use = 0.0075 µg/µL x 20 µL = 1.5µg

➤ Total volume of cell suspension = 500 µL

➤ Mass of plasmid in suspension $= \dfrac{100\,\mu L}{500\,\mu L} = 0.2 \times 0.15\,\mu g = 0.03\,\mu g$

➤ Number of colonies per µg of plasmid $= \dfrac{50\ \text{colonies}}{0.03\,\mu g} = 1.6 \times 10^2$

LAB 9 Biotechnology: Restriction Enzyme Analysis of DNA

Analysis Question: How can we use genetic information to profile individuals?

Exercise 9A: Restriction Enzyme Cleavage of DNA and Electrophoresis

HindIII

Actual bp	Measured Distance in cm
21,130	3.0
9,416	3.9
6,557	4.8
4,361	6.1
2,322	9.1
2,027	9.6
570	Cannot see on gel
125	Cannot see on gel

EcoRI

Band	Measured Distance in cm	Actual bp	Interpolated bp from Graph
1	2.8	21,226	19,000
2	4.4	7,421	9,000
3	4.9	5,804	7,000
4	5.1	5,643	6,800
5	5.7	4,878	5,000
6	6.9	3,530	4,300

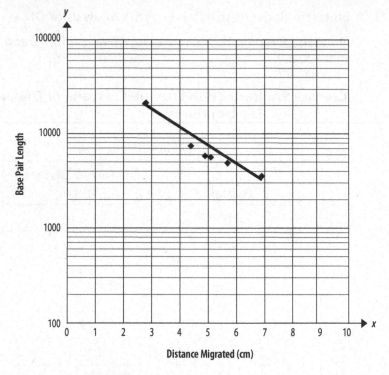

Interpretation of this Exercise

➤ Lambda phage DNA was incubated with restriction enzymes HindIII and EcoRI separately. The migration distance of the DNA bands produced by HindIII were measured in centimeters and were plotted against bare paper size using semi-log paper. This was accomplished with DNA gel electrophoresis.

➤ Drawing the line of best fits allows for interpolation of the same DNA cut with EcoRI. Based on the line of best fit, the base pairs of the lambda DNA can be found and compared to the known value.

Important Points of this Laboratory

➤ Smaller pieces of DNA migrated faster and therefore are farther on the gel.

➤ The electrical current running through the buffer separates the DNA based on size. DNA is negatively charged; therefore, it migrates toward the positive end.

➤ If the restriction enzymes recognition site is mutated, the enzyme will not cut the DNA properly. The result will be the incorrect number and size of bands on the gel.

BIG IDEA 4: SYSTEMS INTERACTIONS

LAB 10 Energy Dynamics

Analysis Question: What factors govern energy capture, allocation, storage, and transfer between producers and consumers in a terrestrial ecosystem?

Exercise 10A: Producers and Consumers

Interpretation of this Experiment

➤ In this experiment, brussels sprouts were fed to butterfly larvae and the energy and biomass flows were calculated at 12 days, 15 days, and after 3 days of growth. Most of the mass of the brussels sprout is water, which is an important product for the larvae to consume; therefore, it is important to understand why only fresh brussels sprouts, and not dried ones, must be used in this experiment. Be sure to understand an energy flow diagram and be able to draw one for this experiment.

Exercise 10B: Energy/Biomass Flow from Plant to Butterfly Larvae

Larva age (per 10 larvae)	12 days	15 days	3 days of growth
Wet mass of brussels sprouts	30 g	11 g	19 g consumed
Plant percent biomass (dry/wet)	0.15	0.15	0.15
Plant energy (wet mass x percent biomass x 4.35 kcal)	19.58 kcal	7.5 kcal	10.56 kcal consumed
Plant energy consumed per larvae (plan: energy/10)	0.2 g	1.5 g	1.3 g gained
Wet mass of 10 larvae	0.15	0.15	0.15
Larvae percent biomass (dry-wet)	0.03 kcal	0.15 kcal	0.12 kcal
Energy production per individual (individual wet mass x percent biomass x 5.5 kcal/g)	0.03 kcal	0.15 kcal	0.12 kcal
Dry mass of the frass from 10 larvae	—	0.5 g	0.5 g excreted
Frass mass per individual	—	0.05 g	0.05 g excreted
Frass energy (waste) (frass mass x 4.76 kcal/g)	—	0.25 kcal	0.24 kcal excreted
Respiration estimate (plant energy consumed—frass waste energy production)	—	—	0.88 kcal

Interpretation of this Experiment

➤ According to the above chart, the wet mass of the brussels sprouts decreased over 15 days and eventually the larvae consumed 19 g of the brussels sprouts after growing for 3 days. The plant-energy row demonstrates the transfer of energy from the plant to the larvae: The plant produces the energy for the butterfly to consume it. The plant-energy-consumed row reiterates this finding, and in fact, the wet mass of the larvae themselves increases as they consume the water content of the brussels sprouts. You want to familiarize yourself with this type of data chart and be able to read it and draw conclusions.

LAB 11 Transpiration

Analysis Question: What factors, if any, affect transpiration in plants?

Exercise 11A: Transpiration

Cumulative Water Loss in mL/m²

	Time (minutes)			
Treatment	0	10	20	30
Room	0	1.50	3.20	4.7
Light	0	4.00	8.12	12.13
Fan	0	4.21	8.45	12.30
Mist	0	1.50	2.00	2.33

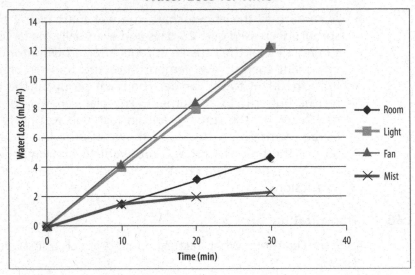

Interpretation of this Exercise

➤ Transpiration, or the uptake of water from the leaf source, is highest with both light and fan conditions. Both of these conditions cause water to be lost from the leaf surface. A water potential is created between the air surrounding the leaf and the photometer where the bottom of the stem contains water. Water will travel from an area of higher water potential to lower water potential. The mist condition mimics increased humidity, decreasing the water potential difference, since more water is occupying the surrounding air. The mist line is actually below that of the control (room) indicating the surrounding air has more water associated with it.

➤ The line graph above reconfirms the data in the chart: Both the fan and light conditions increase transpiration over a 30 minute time period, and in fact, show relatively similar transpiration rates.

➤ The transpiration rate in the room is used as a reference point for the other conditions. For example, the mist condition shows that the transpiration rate is much slower than it would otherwise be in the room; this is also in stark contrast to the light and fan conditions.

➤ Another interesting conclusion that can be made is that the transpiration rate in the room continues to steadily rise over the 30 minute time period; however, the mist condition shows a slight increase of transpiration after 20 minutes had passed.

LAB 12 Fruit Fly Behavior

Analysis Question: What environmental factors affect fruit fly responses?

Exercise 12A: Environmental Factors

Environmental Factor	10 minutes	20 minutes
Salt	0	2
White Vinegar	25	60
Ripened Fruit	28	64
Sugar	10	18
Apple Cider Vinegar and Dish Soap	27	58

Interpretation of this Exercise

➤ The above chart shows how many fruit flies were present around or on the substance after 10 minutes and 20 minutes time. Since fruit flies are attracted to both overly sweet and vinegar-smelling substances, it is no surprise that the most flies were attracted to the white vinegar, the ripened fruit, and the apple cider vinegar with dish soap. The least amount of flies were attracted to the salt, and the flies probably were just checking out the substance briefly. Both the white vinegar and the ripened fruit were similarly attractive to the fruit flies. Although the sugar is obviously sweet, it was dry and crystalized and did not attract as many flies as the "sweet and wet" substances did. The apple cider vinegar with the dish soap also attracted about the same amount of fruit flies as the other vinegar and the ripened fruit; however, this substance also trapped most of the flies in the dish soap, while the apple cider vinegar attracted the flies to their deaths.

Exercise 12B: Reproductive Behavior in Fruit Flies

Interpretation of this Exercise

➤ Many organisms exhibit behaviors that indicate courtship. For *Drosophila melanogaster* a list of male and female characteristics are listed below:

- Male (tend to exhibit behaviors that promote mating): stamping the forefeet, circling the female, and wing vibration.

- Female (tend to exhibit behaviors that do not promote mating): ignoring, depressing wings, or flying.

Exercise 12C: The Life Cycle of *Drosophila*

Interpretation of this Exercise

➤ *Eggs*—small and oval shaped, and usually found on the side of culture tube.

➤ *Larval Stage*—wormlike stage that tunnel through the medium.

➤ *Pupal Stage*—fully mature larva are called pupa and tend to be brown in color. Basic body parts can be observed.

➤ *Adult Stage*—fly emerges from pupal casing and mating can take place again.

Exercise 12D: Crosses

Interpretation of this Exercise

➤ Cross 1 monohybrid
- Assume normal wings is dominant to dumpy (vestigial) wings. Cross a pure breeding long wing (W^+) to a dumpy (vestigial) wing (w).
- F_1 cross: W^+W^+ X ww → All progeny W^+w (all normal wings heterozygotes)
- F_2 cross: W^+w x W^+w → Progeny 1 W^+W^+: 2 W^+w: 1 ww (3 normal wings: 1 short wing)

➤ Cross 2 dihybrid

- Assume gray body color (g^+) and normal wings (w^+) is dominant to black body color (g) and dumpy (vestigial) wings (w). Cross a pure breeding gray and normal wing to black and dumpy (vestigial) wing.

- F_1 cross: $g^+g^+w^+w^+$ X ggww → All progeny $g^+g^+w^+w$ (all gray, long-winged heterozygotes)

- F_2 cross: g^+gW^+w x g^+gW^+w → Progeny g^+gw^+w, ggww, g^+gww, ggw^+w

- (1:1:1:1 gray, normal wings; black, dumpy (vestigial) wings; gray, dumpy (vestigial) wings; black, normal wings)

➤ Cross 3 sex-linked

- Eye color is sex-linked in *Drosophila melanogaster*. Assume red eye (X^{w+}) is dominant to white eye (X^w). Cross pure breeding red eye female to a white eye male.

- F1 cross: $X^{w+}X^{w+}$ x X^wY → All progeny red eye: females are carrier $X^{w+}X^w$ and male $X^{w+}Y$

- F2 cross: $X^{w+}X^w$ x $X^{w+}Y$ → Progeny 1 $X^{w+}X^w$, 1 $X^{w+}X^{w+}$, 1 X^wY, 1 X^wY (all females have red eyes, ½ males have red eyes, and ½ males have white eyes).

Exercise 12E: Chi-Square Analysis

Interpretation of this Exercise

➤ Chi-square is a statistical test to ensure the validity of a hypothesis.

- Null Hypothesis—there is no statistical difference between expected data and observed data.

- Alternative Hypothesis—another hypothesis that explains your observation.

➤ Formula is $$\chi^2 = \sum \left[\frac{(o-e)^2}{e} \right]$$

o = observed number of individuals

e = expected number of individuals

Σ = sum of values

Degrees of freedom = expected phenotypes −1

Use of the Chi-Square Table of Critical Values

Probability (p)	Degrees of Freedom				
	1	2	3	4	5
0.05	3.84	5.99	7.82	9.49	11.1

If the calculated chi-square is greater than or equal to the critical value, the null hypothesis is rejected with a reassurance of 95%, meaning only 5% of the time would you see the null hypothesis as being correct.

Sample Data

Pheno-type	Number Observed	Number Expected	(o – e)	$(o - e)^2$	$(o - e)^2/e$
Normal wing	70	75	–5	25	0.33
Dumpy wing	30	25	5	25	1.00
				$\chi^2 = \sum \left[\dfrac{(o-e)^2}{e} \right]$	1.33

Result: There is no difference between observed and expected phenotypes; accept the null hypothesis.

LAB 13 Enzyme Activity

Analysis Question: How do abiotic and biotic factors influence the rates of enzymatic reactions?

Exercise 13A: Test of Peroxidase Activity

Interpretation of this Exercise

➤ The enzyme is peroxidase with the substrate being hydrogen peroxide (H_2O_2). The products released are water and oxygen gas.

$$\text{Peroxidase} + H_2O_2 \rightarrow 2H_2O_2 + O_2 \text{(gas)}$$

Note: Oxygen gas is flammable and will reignite a glowing flint.

➤ Note: Peroxide is a toxic by-product of aerobic metabolism.

➤ Abiotic and biotic factors should affect the efficiency of this reaction.

Exercise 13B: Determining How pH Affects Enzymatic Activity

pH	3	5	6	7	8	10
	−0.002	0.543	0.321	0.160	0.056	0.004

Interpretation of this Experiment

➤ The above chart shows how pH affects the enzymatic activity of peroxidase. Its optimal pH environment, according to the chart, is around pH 5.0, with also decent activity continuing at pH 6.0. There is almost no activity at pH 7.0 and extremely little activity at pH 8.0 and pH 10.

Exercise 13C: Determining How Temperature Affects Enzymatic Activity

Temp	4°C	15°C	25°C	43°C	55°C	70°C	100°C
	0.102	0.163	0.234	0.308	0.274	0.156	0

Interpretation of this Experiment

➤ The above chart shows how temperature affects the enzymatic activity of peroxidase. Its optimal enzymatic activity is around 43 degrees Celsius, with good activity continuing at 25 and 55 degrees Celsius. The far ends of the activity spectrum that still include enzymatic activity are 15 and 70 degrees Celsius, with an even lessened activity at 4 degrees Celsius. The only absence of enzymatic activity occurred at 100 degrees Celsius because the enzyme became denatured and is unable to function properly.

AP® Biology Equations and Formulas

Statistical Analysis and Probability

Mean

$$\bar{x} = \frac{1}{n}\sum_{i=1}^{n} x_i$$

Standard Deviation

$$s = \sqrt{\frac{\sum(x_1 - \bar{x})^2}{n-1}}$$

Standard Error of the Mean

$$SE_{\bar{x}} = \frac{s}{\sqrt{n}}$$

Chi-Square

$$\chi^2 = \sum \frac{(o-e)^2}{e}$$

Chi-Square Table

p value	Degrees of Freedom							
	1	2	3	4	5	6	7	8
0.05	3.84	5.99	7.81	9.49	11.07	12.59	14.07	15.51
0.01	6.63	9.21	11.34	13.28	15.09	16.81	18.48	20.09

$\bar{x}$ = sample mean

n = sample size

s = sample standard deviation (i.e., the sample-based estimate of the standard deviation of the population)

o = observed results

e = expected results

Σ = sum of all

Degrees of freedom are equal to the number of distinct possible outcomes minus one.

Laws of Probability

If A and B are mutually exclusive, then:

$$P(\text{A or B}) = P(\text{A}) + P(\text{B})$$

If A and B are independent, then:

$$P(\text{A and B}) = P(\text{A}) \times P(\text{B})$$

Hardy-Weinberg Equations

$p^2 + 2pq + q^2 = 1$ $\qquad$ p = frequency of allele 1 in a population

$p + q = 1$ $\qquad$ q = frequency of allele 2 in a population

Metric Prefixes

Factor	Prefix	Symbol
10^9	giga	G
10^6	mega	M
10^3	kilo	k
10^{-1}	deci	d
10^{-2}	centi	c
10^{-3}	milli	m
10^{-6}	micro	μ
10^{-9}	nano	n
10^{-12}	pico	p

Mode = value that occurs most frequently in a data set

Median = middle value that separates the greater and lesser halves of a data set

Mean = sum of all data points divided by number of data points

Range = value obtained by subtracting the smallest observation (sample minimum) from the greatest (sample maximum)

Adapted from College Board AP® Biology Course and Exam Description. Copyright © 2020 by the College Board.

Rate and Growth

Rate

$$\frac{dY}{dt}$$

Population Growth

$$\frac{dN}{dt} = B - D$$

Exponential Growth

$$\frac{dN}{dt} = r_{max}N$$

Logistic Growth

$$\frac{dN}{dt} - r_{max}N\left(\frac{K - N}{K}\right)$$

dY = amount of change

dt = change in time

B = birth rate

D = death rate

N = population size

K = carrying capacity

r_{max} = maximum per capita growth rate of population

Simpson's Diversity Index

Diversity Index = $1 - \sum\left(\dfrac{n}{N}\right)^2$

n = total number of organisms of a particular species

N = total number of organisms of all species

Water Potential (Ψ)

$$\Psi = \Psi_P + \Psi_S$$

Ψ_P = pressure potential

Ψ_S = solute potential

The water potential will be equal to the solute potential of a solution in an open container because the pressure potential of the solution in an open container is zero.

The Solute Potential of a Solution

$$\Psi_S = -iCRT$$

i = ionization constant (1.0 for sucrose because sucrose does not ionize in water)

C = molar concentration

R = pressure constant ($R = 0.0831$ liter bars/mole K)

T = temperature in Kelvin (°C + 273)

$$pH = -\log[H^+]$$

Surface Area and Volume

Surface Area of a Sphere

$SA = 4\pi r^2$

Surface Area of a Rectangular Solid

$SA = 2lh + 2lw + 2wh$

Surface Area of a Cylinder

$SA = 2\pi rh + 2\pi r^2$

Surface Area of a Cube

$SA = 6s^2$

Volume of a Sphere

$V = \dfrac{4}{3}\pi r^3$

Volume of a Rectangular Solid

$V = lwh$

Volume of a Cylinder

$V = \pi r^2 h$

Volume of a Cube

$V = s^3$

r = radius

l = length

h = height

w = width

s = length of one side of a cube

SA = surface area

V = volume

Notes

Notes

Notes

Notes

Notes

Notes

Notes

Notes

Notes

Notes

Notes